How to Eat

MARK BITTMAN is the author of more than twenty acclaimed books, including the *How to Cook Everything* series. He wrote for *The New York Times* for more than two decades, and became the country's first food-focused op-ed columnist for a major news publication. He has hosted two television series and been featured in two others, including the Emmy-winning *Years of Living Dangerously*. Bittman is currently the Special Advisor on Food Policy at Columbia's Mailman School of Public Health, and the editor-in-chief of *Heated*.

DR DAVID L. KATZ is the founding director of Yale University's Yale-Griffin Prevention Research Center and founder/president of the True Health Initiative. He is a globally recognized expert on nutrition, weight management, and the prevention of chronic disease and has written for major media outlets including *The Huffington Post*, *Forbes*, and *U.S. News & World Report*.

How to Eat

ALL YOUR FOOD AND DIET QUESTIONS ANSWERED

MARK BITTMAN
DR DAVID L. KATZ

SCRIBE

Melbourne • London

Scribe Publications
2 John Street, Clerkenwell, London, WC1N 2ES, United Kingdom
18–20 Edward St, Brunswick, Victoria 3056, Australia

This edition published by arrangement with Houghton Mifflin Harcourt

Published by Scribe 2020

Book design by Melissa Lotfy

Printed and bound in the UK by CPI Group (UK) Ltd, Croydon CR0 4YY

Scribe Publications is committed to the sustainable use of natural resources and
the use of paper products made responsibly from those resources.

9781913348267 (UK edition)
9781922310170 (Australian edition)
9781925938340 (e-book)

Catalogue records for this book are available from the National Library of
Australia and the British Library.

scribepublications.co.uk
scribepublications.com.au

Contents

Introduction

Science, Sense, and Mashed Banana

Let's say you feel like eating a banana. You'd likely peel it first.

You might have the world's most powerful food processor at your disposal, but that would not help you peel your banana any better. You might have an even more powerful chainsaw in your garage, or a table saw. But these wouldn't help either.

You know that using any of these would be ridiculous. However state-of-the-art each tool might be, however representative of the prowess of the applied science of engineering, none would enhance your native banana-peeling capacity. In fact, each would leave you with a mangled mess to clean up where once had been your banana.

Science is a set of power tools. Used well, the tools of science (which, of course, is the incubator of all other power tools) extend human ability and perception. With a microscope, a telescope, or a Gauss meter, science can reveal what was previously too small, too distant, or, in the case of an electromagnetic field, imperceptible. Science enables us to see the invisible, hear the inaudible, discern the elusive, and understand the otherwise inscrutable.

But science is not *only* about power tools, and one of the great misdirections in modern nutrition is the contention that to derive truth and understanding you must always only use a particular power tool, a special kind of scientific study, and that without this kind of study nothing can be understood. Our response to that is: nonsense.

To be clear, we are both disciples of science. One of us

(David) has made a career of it, running a research lab, conducting and publishing randomized trials, systematic reviews, and meta-analyses, and even inventing methods of evidence synthesis — methods for reaching conclusions about the weight of evidence on a given topic — along the way. The other (Mark) became entranced by science as a boy, but chose another direction, one that routinely relies on science as a source of understanding.

But we are equally adamant defenders of *sense*. Science without sense is the chainsaw approach to peeling a banana. You don't always need a power tool. Sometimes it's not only superfluous but potentially harmful. We don't need studies to tell us how to peel a banana, or that an apple thrown into the air will come back down, or even that the apple is a good thing to eat.

The artful (or at least competent!) blend of science and sense is what we believe to be our signature contribution, the approach that sets *How to Eat* apart from the literally hundreds of "nutrition" books out there. Throughout this book, we invoke sense to interpret science. We rely on sense to discern the readily observable. We employ sense to assess the relevance of science. Science may be the best means ever devised to answer hard questions, but only sense can determine whether you are asking a valid, useful question in the first place, and sense can inform whether science is needed, and if so, what kind, to answer that question.

We each know all we need to know about how to eat for pleasure. We know much about how to eat for health and sustainability from experience and observation, too, and far more from science of every description, from basic science to intervention trials and observational epidemiology in large cohorts and even whole populations.

We increase our knowledge, and can rely on it even more, when we accumulate evidence and use its weight to make judgments, rather than — as the 24-hour news cycle encourages us to — by tilting with every new study, believing that a new finding somehow sets the world off kilter.

We know most and know it best when all sources of genuine insight are conjoined. Sense *plus* science.

Any given research method is just one kind of tool. Used well, it can help build the bridge to truth and construct understanding. Used badly, it will just mangle your banana. The insistence that nothing is "true" unless and until it's confirmed by a randomized controlled trial is not only wrong — and you don't need to be a scientist to realize this, nor should you let a scientist convince you otherwise — but is often used by people with ulterior motives who seek to profit by our confusion. We could design a trial, for example, that would conclude that overeating sugar is better than overeating fat — or vice versa — and yet sense tells everyone that overeating either is not a good thing.

Ultimately, we really just want to know what is true; we want to understand. We expect you do, too, and using science and sense, we attempt here to tell it like it is, and we tell you *how we know* how it is. And when we aren't sure, we tell you that; not everything is known. Both science and sense leave room for doubt, and often require it.

But beyond the shadow of all doubt, and resting securely on a foundation of both science and sense, is a sufficient understanding of how to eat to massively reduce your risk of all major chronic disease and premature death. We have, and can give you, the understanding necessary to add years to your life and life to your years, and to help save our existence on this planet.

No, we don't know everything. But we know *enough*. Science, through the filter of sense, reveals more than enough reliable truth about how to eat to do a literal world of good.

—DK/MB

QUESTIONING THE QUESTIONS

(Why Do We Even Need to
Ask How to Eat?)

How Did We Get Here?

Shouldn't "how to eat" be clear already?
In a way, it is. Every animal knows how to eat, and only in humans (and the animals under their control) has this been perverted. In the last century or so, we've been led astray. And we're so far from our origins that it's proving hard to find our way back.

Yet there's nothing more important: Food is the fuel that runs every function of the complex human machine. It is the source of construction material for the growing bodies of children and all the replacements adult bodies require on a daily basis. What we eat is crucial to the integrity of the nervous system, the balance of hormones, the function of our blood vessels, the responses of the immune system. If there were one thing we'd say, it's this: You don't want to eat non-food. Period. But we'll get back to that.

Take me back to a time before mass food production.
Close your eyes and try to imagine: Nearly every culture since the dawn of agriculture relied on grains as the foundation of their diet; they were (and are) inexpensive because they were (and are) easy to grow in quantity.

Everyone ate mostly . . . what?
Virtually everyone ate grains: rice, wheat, corn, millet, depending on the region. Grains were staples. In some places, the aristocracy ate delicacies.

Like what?
Like meat. Until recently, eating meat was a rare indulgence.
Which wasn't a bad thing: Low consumption of animal prod-
ucts is good for all parties concerned (humans and animals).
And until the twentieth century the harm imposed on the an-
imals was minimal.

The twentieth century was a turning point, and a marked
departure from all prior history. While people used to eat food
that was almost exclusively grown nearby, now we have com-
modities — including meat — produced in industrial conditions and shipped all around the world. And the harm imposed on animals raised for food is incalculable.

**But to be clear, until recently humans lived on a plant-based
diet?**
Yes, almost entirely; there was no choice. Where there are
exceptions due to survival imperatives — where people like
the Inuit must eat a great deal of meat because that's what
there is — they are not generally associated with enviable
health outcomes. But most humans were plant-predominant
omnivores. We have a really good extant example of that in
the Bolivian Tsimané — also known as the Chimane — a
tribe of modern-day foragers who came to scientific attention
in the last couple of years because they have the cleanest coronary arteries known to science, and have absolutely no heart
disease.

How is that possible?
A big part of that is their lifestyle, which includes a mostly
plant-based diet. They eat plants that they grow and find locally. If it's a bad crop season, they hunt and fish.

It's not just a connected, healthy way to live — it's their
only choice. Throughout history, the traditional practice of
eating a plant-based diet was about expediency, survival, and

making the most out of what was available; people weren't thinking about environmental footprint or ethics. Those are modern indulgences: We have so much food — and so much impact — that we must make choices.

How do we even know that in the past, almost everyone was a plant-predominant omnivore?
Evolutionary biology. We also know what there was to eat, and we know that humans are naturally omnivorous. What our taste buds like is not accidental. What tastes good is associated with survival.

Then why do I adore ice cream? That can't be the best choice for my survival.
We like sugar for good reason — breast milk is sweet; so are fruits.

What's the reasoning behind my addiction to salty food?
We can think of our origins in the briny deep where sodium is everywhere; we were adapted to actually be soaking in the stuff. Then we crawled out of the ocean and figured out how to make a living on land — leaving all of our treasured sodium behind. We were obligated to go from being good at expelling the excess (as all sea creatures do) to being good at finding some (as most land animals work to do).

In the natural world on land, it isn't easy to get a lot of salt, which is why most terrestrial animals look for it. (This is why your dog likes to lick your skin when you're sweaty, deer go to a salt lick, elephants gather to suck up mineral-rich mud, and so on.) Once humans began to mine salt, it became easy to get. But our cravings for it are ancient, and baked into our DNA: Give us a more-than-adequate sup-

You don't stop having Stone Age impulses just because they no longer serve you well in a modern world.

ply, and we may overconsume. You don't stop having Stone Age impulses just because they no longer serve you well in a modern world.

Can evolutionary biology explain why I love fried food and a greasy hamburger? It would make me feel better to have scientific justification.

In addition to responding to sweet and salty tastes, our taste buds also reward us for the feel and texture of fat because it's the most energy dense of the three macronutrients, with more than twice as many calories per gram as protein and carbs. If you were living in a world with scarce food and working hard to chase after your calories, fat would be a huge win. So, yes — it makes sense to love grease, to love salt, to love sweet; they're the trifecta of survival in a world where finding food meant hunting and foraging.

Why aren't our taste buds designed to find vegetables as delicious as junk food?

Plants tend to be energy dilute; wild animals aren't high in fat, either. People had to work really hard to find high-calorie, energy-dense foods like nuts, seeds, eggs, organ meats, and bone marrow. It makes sense that the people who found those foods were better at making babies than the people who didn't.

You're saying that all the unhealthy cravings I have are linked to ancient survival instincts?

Yup. But remember, these things were not unhealthy when they were scarce and hard to find; rather, *our ancestors needed them.* Salt is not unhealthy unless you get too much, and you need some to live; same with fat. Sugar is not unhealthy when the only places to find it

One of the most potent determinants of dietary preference is familiarity.

are breast milk, whole fruits, and honey you have to wrestle from a swarm of angry bees. The cravings were good in their native context; we've changed that context.

Ha! That's ironic.
Yes, and unfortunate: Food processors take advantage of our taste buds to make us buy their products.

Are we so easily manipulated? Is there a way out? Or are we doomed to eat junk forever? The good news is that taste buds are adaptable little things (they're bundles of nerves, actually), and readily learn to love the foods they're with. One of the most potent determinants of dietary preference is familiarity. So, yes, we all have some native tendencies cooked into our DNA. But what we choose to eat powerfully shapes our actual preferences.

Consider the transcultural evidence of that: Traditionally, Mexican babies learned to love rice and beans; Indian babies learned to love dal and chapati; Japanese babies learned to love fish and tofu; Inuit babies learned to love seal; and American babies . . . learned to love Froot Loops. This shows that what we eat can shape what foods we prefer. And in that resides the tremendous promise of "taste bud rehab," and the chance we all have to learn to love food that loves us back.

I need to start my taste bud rehab ASAP. There are so many diets I could choose for this. If you had to use one word to describe a healthy diet, what would it be?
Balance.

Like "everything in moderation"?
Corny as it is, that's about the right mantra for diet. The Mediterranean diet, for example, is high in monounsaturated and polyunsaturated fats, and contains some saturated fat. But you

can't point to saturated fat in this instance and say that it's doing any harm, because it fits into an overall balance of nutrients. Balance confers an overall health benefit.

But let's be careful about "everything" and "moderation." There *are* foods — let's call them junk foods! — that are best avoided altogether, for reasons of health, ethics, and/or environmental impact. And moderation readily becomes a slippery slope, where a little of this and a little of that . . . adds up to a LOT of "this + that." But if a diet is mostly made up of the right stuff — vegetables, fruits, whole grains, nuts and seeds, beans and lentils, and plain water — then truly most else in genuine moderation would be okay.

ANSWERING THE QUESTIONS

(All About Eating and Health
—a Multicourse Q&A)

What Is the Best Diet?

Can we define "diet," please?
The term "diet" has been converted into a pop culture catch-phrase. It's something that you "go on" to lose weight, a short-term (non-) "solution." But "diet" comes from the Latin for (we're translating loosely here) "lifestyle," or daily food intake; it's not a way to eat to lose weight as fast as possible (and gain it back even faster.) It's how you eat for life. It's a thing you do to remain healthy. So you want a good diet, permanently. You don't want a two-week "diet." And there's almost nothing more important to say about the word than that.

Where do diets come from?
Diets are ways of living that groups of people practice for generations. That's fundamentally different from the notion of a renegade genius who says: "I came up with a new way to eat, and I'm going to sell it to the world even though there's no long-term proof of benefit or even safety."

So the cabbage soup diet, the grapefruit diet, the Hollywood cookie diet . . . ?
Perfect, ridiculous examples — yes. There are no populations of people who live on cabbage soup only, so it's preposterous!

I mean, every other day there's a new diet craze promoting weight loss, mental clarity, radiant skin — basically near human perfection. How can people make these claims? Is there actual evidence?
There is lots of evidence, and the weight of it tells a clear,

consistent story. But there are also many opinions competing for attention, opinions that — intentionally or not — sow confusion.

Science was never designed to work well with news cycles; there isn't a punch line every twenty minutes. Science is cumulative, incremental, developing over months, years, decades, centuries. It takes time to approach truth and understanding.

Trying to shape scientific discovery into short news cycles for the sake of views, clicks, book sales — whatever — turns it into a pretzel; it leads nowhere. There really *isn't* confusion when science is used right; there's pseudo-confusion when it's used badly.

Science is a tool, and any tool can be used well or badly; a hammer is great for nails, and just as horrible for screws. Science is a power tool, for populating gaps in our understanding. But used badly, it can mangle understanding instead.

Bottom line: There *is* evidence of the impact of diet on human health — good, clear evidence, indisputable and mostly uncontroversial. We know what a good diet is. We know how to practice it.

Then why do so many diets—like the Whole30 diet or ketogenic diet—literally ask us to eat in an imbalanced, highly limited way?

Those are not diets for life; they're ostensibly short-term weight loss diets, though even that's arguable. Beyond that, they're simply not good choices: Balance is good, imbalance is bad. Really. Period.

A lot of things not consistent with balance or health can lead to rapid weight loss in the short term — a bout of flu, for instance, or for that matter, cholera! Gimmicky weight loss diets substitute severe restrictions for long-term healthful eating. They work for quick weight loss, but they're not sustainable.

Balance, on the other hand, is a high-level principle that pertains across all considerations of diet and nutrients. For example, you need sodium to live; you just don't need as much of it as modern, highly processed diets deliver.

> In an ideal world, all ultraprocessed foods would be eliminated.

But aren't some nutrients, like saturated fat, just evil, to be avoided at all costs? In some cases, isn't elimination more important than balance?

In an ideal world, all ultraprocessed foods would be eliminated, because food would actually once again be ... food. In that world, there would be many, many fewer bad choices to make. If we're talking about naturally occurring nutrients, though, things like saturated fat are not intrinsically "bad." Nutrients like saturated fat or sodium are only "bad" because modern diets provide too much of them relative to other sources of nourishment. That's imbalance. Some saturated fat is found in even the most ideal diets; and sodium is an essential nutrient. As Paracelsus, the father of toxicology, famously said: The dose makes the poison.

Is there a "diet" that leaves all the others in the dust?

It would be truer to say that we know *eating patterns* that beat out other diets — but as soon as we move in that sensible, defensible direction, all the pixie dust drops out of the equation; it doesn't sound like magic. Sadly, most people are convinced they want pixie dust, no matter how many times false promises about its magical powers have let them down, and no matter how simple good eating is shown to be. Whenever we're comparing contemporary diets, from intermittent fasting to

Whole30, there are commercial interests attached. But *the simple truth is that all "good" diets share the same principles.* They're variations on a theme.

What's the theme?
The best diets have this in common: They focus on foods that are close to nature, minimally processed, and plant predominant — what we call a whole-food, plant-predominant diet. Everything else is detail.

How do we know that plant-predominant diets are the healthiest?
What we *don't* have is a single randomized trial, beginning before birth, lasting a lifetime, enrolling tens of thousands — to show once and for all "what diet is best." What we do have is a mountain of evidence, built a bit at a time, supporting the theme of wholesome foods, mostly plants, in balanced, sensible assemblies.

We can't say what "diet" is best. We *can* say what eating patterns are best. And that's it: real foods, close to natural form, mostly plants, augmented a little bit by almost whatever else you like. It's that simple, as hard as that might be to hear.

What nutrients am I guaranteed with a whole-food, plant-predominant diet?
Everything you need. You are getting a wide variety of vitamins, minerals, and antioxidants from all those plants, and omega-3 fatty acids from nuts, seeds, and seafood. When most of your fats are from plant sources and fish, almost all of those fats are polyunsaturated or monounsaturated predominant (see page 160 for more about fats). You're getting the highest quality carbs, with plenty of fiber, and you're getting plenty of protein. You're even getting peace of mind! There could still be minor nutrient gaps — vitamin D, for instance, or vitamin B12

— depending on your specific choices. These are readily addressed with supplements or minor changes in diet or lifestyle.

Okay, so I just have to work on eating close to nature, avoiding ultraprocessed foods, and eating a lot of fruits and veggies—that seems pretty doable! Now, should I be eating high or low in fat? What about high or low in carbs?
It doesn't matter.

What?!
You don't need to worry about it. One of the greatest distractions in modern thinking about diets is how high or low a dietary pattern is in particular macronutrients. You can do what you want. Don't see that as a problem — see it as liberating!

Well, there's got to be more to it than that.
Yes. You can eat a diet that's high or low in fat or carbs, but there are good and bad ways to do either. A high-fat diet could mean cream puffs and pepperoni; or it could mean walnuts, avocado, olive oil, and salmon. A low-fat diet can be Coca-Cola and cotton candy — in fact, they're both fat-free! — or it can be mostly fruits and vegetables. The same is true of carbs: A vegan diet is usually a high-carb diet; so is a junk food diet. The important thing to remember is that any diet rich in whole vegetables, fruits, beans, lentils, nuts, seeds, and whole grains with relatively little of all else is a good diet. And yes, that can mean a high-fat diet, as long as that fat is "good" fat.

As we'll discuss on page 155, it's actually pretty hard to eat truly "low carb" — all plant foods are mostly carbohydrates — and eat well. But a good, high-fat, and adequate-protein diet can be relatively low-carb, and healthy. It'll still be mostly plant-based, with the fat coming from good oils and the protein coming from beans. The highlighted sentence on page 15 remains key. Whether it's high-carb or high-fat or otherwise is not as important as whether it's a mostly plant-based diet with little of all else.

What do good diets look like in practice?
Great question. To expand on what we've said on the last two pages, here's the fundamental statement: Diets of the world's longest-lived people are rich in veggies, fruits, beans, lentils, whole grains, nuts, seeds, and good fats like olive oil or fat from fish. They de-emphasize meat and dairy and exclude ultraprocessed foods and soda.

A good diet is also one that really works for populations, meaning it works over time, and that whole families can do. If you want a name for it, you can call it "Mediterranean" — that's the most popular example — but there are many variations and they're all good. We'll quote Michael Pollan here: "Eat food, not too much, mostly plants." Food means things that exist in nature or close to nature. Froot Loops and Coke, for example, are technically plant-based, but they're not food.

What do you mean when you say that a diet "really works"?
That it leads to good health, longevity, and vitality.

Besides the Mediterranean diet, what examples do we have of diets that work for whole populations?
Some of the best known examples these days are from what are sometimes called Blue Zones. These five very diverse populations from around the globe have the highest concentrations of people living to one hundred, free of chronic disease. Their diets range from high-fat to low, vegan to vegetarian to omnivorous — but all are variations on the same basic

theme: wholesome foods, mostly plants, in balanced, sensible, time-honored assemblies. We're going to start to sound repetitive here, but that's the bottom line.

What are these Blue Zone people doing right?
The question of why Blue Zone people live so long led to an assessment of their lifestyle and cultural patterns. These are all groups that have managed, so far, to hang on to traditional lifestyle practices despite the influences of the modern world due to cultural priorities or geographic isolation, or a bit of both. Think of them as "islands of yesterday" where people eat the way their grandparents ate, not the way a fad diet author told them to eat! They're also places where neighbors all know and support one another. They just happen to be far-flung, in Ikaria, Greece; Sardinia, Italy; Okinawa, Japan; Loma Linda, California (it's a traditional Seventh-day Adventist community, and therefore vegetarian); and Nicoya, Costa Rica. Maybe there are more out there somewhere, waiting to be discovered. Also possible is that these "islands" will be drowned by modern intrusions — technology and junk food — and sink beneath the waves of history.

Why do the Blue Zone diets all have a similar reliance on plant foods?
Actually, grains and produce were staples in *all* of the world's diets until we achieved mass production of ultraprocessed food.

Eating a wholesome, mostly plant-based diet sounds like such a moderate request in comparison to what I thought my options were, like drinking only green juice for a week. Since these rules are pretty loosey-goosey, I'm curious: What sets different variations of plant-based diets apart?
Idiosyncrasies. One of the idiosyncrasies of the Mediterranean

diet is that it is high-fat, whereas the traditional Okinawan diet is low in fat. But both follow the rules described above: high in veggies, good fats, little or no ultraprocessed food, etc.

What determines these idiosyncrasies?
Geography and culture. History and tradition. All the best diets emphasize grains, but the traditional grains in Asia tend to be rice and millet; corn, quinoa, and amaranth in Central and South America; wheat and barley in the Mediterranean; and millet and teff in Africa. Not exclusively, but that gives you an idea of the variety. And it doesn't seem to matter. Yes, quinoa is higher in protein than rice, but that doesn't seem to matter, either.

The specific grains you have access to, and for that matter the specific fruits and vegetables, will vary by the part of the world you live in. They are more alike than different.

So whole grains are another major indicator of a healthy diet?
Whole grains are a fixture in all of the world's best diets, just as fruits and vegetables are a fixture.

Does it matter what type of grain I'm eating? There are so many options, and I want to make sure I'm choosing the healthiest grain.
Set aside that worry: None of these idiosyncrasies matter much. You can have a good diet with barley or quinoa, millet or oats. The specific grains you have access to, and for that matter the specific fruits and vegetables, will vary by the part of the world you live in. They are more alike than different.

What matters is that the diet is plant-based?
Yes. That's the part that's nonnegotiable. That doesn't mean necessarily vegan; it means necessarily "near vegan." You want

plenty of vegetables, fruits, whole grains, legumes, nuts, and seeds. That is where the Mediterranean and Okinawan diets, for example, overlap.

Just because the Okinawan and Mediterranean diets work for populations of people, does that guarantee that they will work for me?

Well, nothing about health is ever "guaranteed" for any of us. But for the most part, and for the most fundamental of reasons, yes: You are human! We are all the same kind of animal. Consider that for every other kind of animal, we don't think much about which diet is best for each individual or even group; we think about the fundamentals that are good for the species. The best diets for humans reflect those fundamentals that are good for all of us.

So, yes: They will work. Both the Mediterranean and Okinawan diets pass the "does it work in the real world?" test, for entire populations over extended periods of time. Again, neither is a "diet" in the contemporary sense; they're ways of eating long term, so they're diets in the traditional sense.

The best diets for humans reflect those fundamentals that are good for all of us.

Any closely related variant would work, too. Now, which variant is the best of the best for you — based on taste preference, convenience, and maybe even your own idiosyncrasies like allergies and sensitivities and specific genes — is something only you can answer. But stick with the theme, and you'll never go too far wrong.

There's been an attempt to rebrand some of these diets into promises of weight loss and health. "The Okinawan Diet," for example.

Yes. Keep in mind, though, that when populations do this right, they do it right their whole lives; they probably don't need to lose weight! We've invented diets for weight loss because ... industry invented foods that led to diets that made us fat. (This is literally true: The food industry engineers foods that maximize the eating required to feel full, and they do so for the most obvious of motivations — profit.)

We register made-up diets as contestants in the beauty pageant we are obsessed with watching. But again, *the best diets are traditional ways of living.* Take a fully plant-based diet — you can call it veganism: It's healthy, it's kind to other creatures, it has minimal impact on the environment. When you consider that trifecta of priorities, it's really hard to argue against a vegan diet of real foods, or some close approximation of it.

Should I personalize what I eat for my genetics?

That's nutrigenomics, a kind of personalized nutrition. But Stanford University's most recent study of personalized nutrition found that despite assigning diets that would hypothetically be best for certain genetic codes, they all did exactly the same: The diets were essentially indistinguishable. No matter what genes you have, you were equally likely to lose weight on a good low-fat diet and a good low-carb diet.

So what's a better concept of personalized nutrition, then?

It's "You're in charge." Personalize your dietary patterns in terms of your schedule, routine, and general preferences — within the realm of healthy options, of course. But now you know how simple that is.

ON WEIGHT LOSS

Is there an ideal weight?

Like so much about diet, it's intriguing that we have managed to become confused about things that are obvious. There's no single ideal weight for everybody, but there is a healthy range of weights.

Why should that be obvious?

Let's say we were talking about squirrels. Their body size shifts a little bit with the seasons: They gain weight when they can, and they lose weight when they don't have enough to eat. But squirrels, like all animals, have a particular body size and shape that's suitable for living as a squirrel.

What does the weight of squirrels have to do with humans?

That same sort of thinking pertains to all creatures. There is a range of body composition that is related to the functions of survival.

It's great that squirrels have bodies that allow them to jump from tree branch to tree branch, but what is the natural body size and shape for humans?

In the past, humans were high-stamina foragers. It's harder to traverse lots of miles and gather food and hunt for food if you're carrying a lot of extra weight. So historically and pre-historically, there was a native weight range for people. People are meant to be lean, or at least lean-ish. That was not a challenge throughout most of human history; the challenge would have been finding the surplus calories needed to gain weight. So, we are adapted to be lean. But we're also adapted to eat palatable food whenever we find it. Since few of us are foragers, and most of us are entirely disconnected from natural sources of food, we no longer use our bodies the way our ancestors used them and we have as much food as we can possibly eat at our disposal. That combination is the source of a great deal of modern ill health and the perennial preoccupation with the challenge of weight control.

Our bodies' systems are calibrated to perform well within

a given weight range; when we go outside that range, our bodies may start to fall apart. We're sympathetic to the arguments for health at any size, and adamantly opposed to weight bias — but the fact is that the human body tends to stay healthy when near to the conditions for which it is adapted, and to deteriorate when it wanders far from those. Weight is part of this story — and excess body weight routinely and robustly conspires against optimal health. That doesn't justify weight bias; it's just a fact of epidemiology.

Out of the ideal weight range both in terms of being underweight and overweight?
Yes. When there's too little lean body mass from disease or undereating, it can take you outside of that range. Too much body fat or too little lean body mass are both health threats. We just happen to see more of the former, mostly because there's no profit in un-selling food. Food companies want us to eat as much as possible.

How does your body start to malfunction when your weight is outside those native levels?
Starvation — and loss of lean body mass — will start to turn off growth and repair functions, turn off fertility, impair immunity, and eventually cause everything to shut down.

But of course, the prevailing problem in our culture tends to be excess weight, and excess body fat in particular. That tends to produce a variety of metabolic derangements such as changes in insulin release; changes in the microbiome; even changes in gene expression (how a gene, and the levers and switches in chromosomes that regulate genes, translate into specific levels of particular compounds in the bloodstream); stresses on the heart and blood vessels; hormonal imbalances; increased inflammatory activity; and higher blood pressure. And that's a short list.

Why do people store excess fat in different places?

Because we are all a whole lot alike (as with all human variations), but also a bit different in important ways. Those ways include sex: Men tend to store fat around the middle; women (before menopause especially), much more so in the lower extremities. We also vary genetically: Some people are more prone to store excess calories here; others, there. These genetic differences likely derive, in turn, from adaptations over many generations to a diversity of environments, diets, and conditions. The genes that tend to cause weight gain around the middle, for example, may also be associated with extreme "fuel efficiency" — so they turn up in those of us whose ancestors had the hardest time finding enough to eat. Where fat is distributed in the body makes a major difference in whether or not it causes overt metabolic problems. Interestingly, there are two predominant patterns, known as apple shape and pear shape, or android and gynoid.

Is there a reason why men and women gain weight in different places?

The immediate reason is sex hormones. The root reason — and this is true about almost everything in biology — is adaptation and evolutionary biology. Women, especially pre-menopausal women or fertile women, may have a particular need to be able to store body fat in a safe way.

Why do women need to store body fat?

That need is the obvious one: They need the capacity to feed two rather than one. If we think in terms of adaptation, evolutionary biology, and survival with limited food sources, women who had stored fat were more likely to have bodies capable of reproduction. When women's body fat falls below a particular threshold, it's common for them to become amenorrheic (i.e., stop having their period). That's adaptive.

Essentially, women can't become pregnant when they don't have a sufficient amount of energy reserve available. Related to that, the range of body fat that is healthy for women is about twice the level that's healthy for men.

Because it would be unsafe to get pregnant if you don't have a big enough energy reserve.
Exactly. If you barely have enough energy stored to make sure you're okay — add an extra body to the mix, and everybody's in trouble.

So in women, fertility is actually linked to whether they have an adequate amount of body fat?
Correct. That, in turn, suggests that there would be a survival advantage to women being able to store fat safely. So that's why women have a higher normal body fat percentage than men, and why it generally takes women longer to lose weight.

When women reach menopause, the influence of the sex hormones goes away and women become much more subject to gaining fat around the middle. The risk of heart disease goes up for women after menopause for a number of reasons, but partly because of the effects of menopause on weight gain and weight distribution.

I didn't realize there were better and worse ways to gain weight.
The apple/android weight pattern is the way men tend to gain weight: around the middle. That's the bad metabolic actor. Women tend to gain weight in the pear/gynoid pattern — in the hips, thighs, buttocks, and legs — which is less associated with health problems because it has lower negative metabolic effects. As a rule, fat stored in the lower extremities doesn't tend to drive up blood pressure or lipids, cause insulin resistance, or infiltrate vital organs like the liver.

Why is weight gain around the middle a bad actor, as you say?

Weight accumulated around the middle tends to be rich in adrenergic receptors, which are receptors that respond to stress hormones. The interactions between hormones and these particular fat cells lead to increased inflammation. That inflammation, in turn, is associated with atherosclerosis, increased risk of diabetes, and heart disease.

More important, fat accumulation around the middle tends to mean fat infiltrates the liver, which impairs liver function and causes insulin resistance. That drives up insulin levels, which further favors fat deposition around the middle — so the whole problem is self-propagating. The difference between apples and pears isn't completely reliable in terms of the sex divide, of course; some women tend to gain weight in the android pattern and some men in the gynoid pattern. And, as we've said, women become more vulnerable to the woes of apple-shape weight gain after menopause. But overall, the pattern is pretty consistent, and it makes sense. Because women uniquely bear the burden of storing "fuel for two" in their bodies, it's adaptive for women to be able to store a surplus of body fat without disrupting metabolic balance.

Of note, the same characteristics that make "men's fat" more dangerous make that fat easier to lose. Typically, if a heterosexual pair, like a husband and wife, go on the same diet, he'll lose weight faster than she will. Those receptors in fat around the middle help mobilize it more readily when calories are restricted.

So where weight is stored matters for health.

Enormously. If we are going to talk about an ideal weight, it depends on your sex, your age, and where you're storing fat. There is a great deal more latitude if you're a fertile premenopausal woman storing excess weight in the lower extremities. You may not love that, but it may not be causing you much, or even any, metabolic harm.

As opposed to a middle-aged man who stores it all around the middle.

Where even a little bit of extra weight may be quite danger-ous. In fact, there are whole populations where "weight," per se, can be normal, but where a bit of excess fat around the middle causes metabolic mayhem, often called metabolic syn-drome — a diabetes precursor. This phenomenon is known as lean obesity, because a bit of excess fat in the wrong place pro-duces all the consequences of severe obesity without the ex-tra weight.

So you can experience the negative health effects of being overweight even if you're not?

Technically, you wouldn't be overweight, but you'd be "over-fat." Some people can see significant shifts in their biomarkers — blood glucose, blood insulin, lipids, inflammatory markers, blood pressure — with a weight fluctuation of as little as 3 or 4 pounds. Genetically, some people take whatever little extra fat they produce and put it in all the wrong places, specifically around the vital organs, especially the liver, where, as we note on page 25, it starts to disrupt the processing of insulin, rapidly leads to insulin resistance, and increases a whole array of meta-bolic disruptions that are potentially quite dangerous. If you gain a lot of weight without putting it in the liver — as some people do — you are much less prone to those.

So it's safe to say the relationship between body fat and health is a lot more complicated than weight loss diets make it out to be?

Absolutely. It's way more complicated. Not all body fat is created equal. Some is harmful, some is not. Some is easy to lose, some is very hard.

Is there a way to measure if I'm a healthy size, beyond just weight?

Well, there's body mass index (BMI) — weight in kilograms divided by height in meters squared. A BMI above 25 is considered overweight, and risk for heart disease and diabetes starts to go up from there. Stage one obesity is above 30, stage two is above 35, and stage three is above 40. But because BMI is insensitive to where your weight is — or, for that matter, whether it's mostly fat or mostly muscle — it's a measure that's more useful for seeing population trends than for gauging personal health.

For example, if you just use BMI, elite bodybuilders would register in the obese range, even though most competitive bodybuilders have stunningly low body fat; BMI can't differentiate fat from muscle. But BMI is useful at the population level because we know we have epidemic obesity, not epidemic body building! And since most of us are more prone to excess fat than excess muscle, it can be pretty reliable to say we are going to be significantly better off below 25, substantially better off below 30, and so forth. BMI is also useful for keeping track of our personal weight trends over time.

So what's better than body mass index as an indicator of health related to weight?

Waist size! The best measure related to personal health just uses a tape measure! Waist circumference should be less than 40 inches in men and less than about 34 inches in women to avoid serious risk of insulin resistance.

But remember: When it comes to weight, there is no single, universal cut point. But there is, absolutely, a healthy, normal, adaptive body composition and weight range for *Homo*

> We're built to
> protect ourselves
> against starvation
> way more than to
> protect ourselves
> against obesity.

sapiens — like all creatures — that varies somewhat by gender, ethnicity, age, and so on.

We could just say, "You know if and when you're overweight." The measurements are less important than an honest self-assessment about where you are relative to where you've been, and where your weight seems to be headed. Usually, the formal measures just validate what you already know.

Does body composition change with weight fluctuations over time?

It's important to remember that we're built to protect ourselves against starvation way more than to protect ourselves against obesity; our pre-agricultural ancestors ate as much as they could, whenever they could, because they didn't know when they'd find food again. *We're hardwired to eat all we can.* That's not especially helpful now, when food is ubiquitous and engineered to be hyperpalatable and all but addictive. And it's likely that with repeated cycles of gaining weight and losing weight, the body will gravitate toward the higher weight.

Well, that's super unlucky. Why is that?

There's some evidence that repeated cycles of weight loss and gain make weight loss harder each time. This makes sense in terms of adaptation, and may also be partly due to changes in body composition (i.e., a rising percentage of body fat with such cycles), in the microbiome, and maybe gene expression (see page 22) as well.

Here's one reason why: If you gain weight from overeating, you gain extra fat. That's where those extra calories go. If you lose weight by going on a diet and starving yourself, you will lose both fat and muscle. It's impossible to starve yourself and just lose fat. It doesn't work that way, because your body burns a mixture of fuels.

You can preserve the muscle by working it, and preferentially (but not completely) shift the burn to fat, which is why just going on a diet on its own is not a particularly good idea; it's better to combine exercise with dieting.

Is quick weight loss a bad thing? A grapefruit diet, or fasting, or whatever?

If you just go on a short-term diet as so many people do, you'll lose both fat and muscle. If you then go off the diet and gain weight back, unless you work out like a fiend, you'll gain back mostly fat. With each of those cycles you shift your body composition more and more toward a higher fat percentage, which is a less metabolically efficient machine. Fat requires fewer calories to maintain its size than muscle does. So essentially, you create a pathway by which you need fewer calories each time to maintain fat and require ever more severe calorie restriction to lose it. In other words: Ouch.

That's horrible news. How can I avoid those harmful weight fluctuations?

Taking good care of yourself by eating well and being active consistently, over time, is a much better idea than dieting. How's this? Don't DIEt — LIVEit. Cute, right? But also, true; and important. Lifestyle for finding health is so much better than dieting for losing weight.

That's a very strong argument against yo-yo dieting!

Yes, it is. Absolutely. And that's most people's experience — it gets harder to lose weight each time around.

If it's just going to become harder to lose weight every time, is it even worth the effort to lose weight at all?
Fluctuating in weight may be better than not losing weight at all, and besides, what's the alternative? The best approach is to make a stable adjustment to your dietary pattern and combine it with physical activity that works your muscles rather than thinking, "I'm just gonna starve myself for six weeks and then do it again and again."

If you work toward health, your weight will get "better," wherever it is starting. If you work only to lose weight, however, it may not benefit your health, and could even harm it.

What's the damage of being overweight, really?
Type 2 diabetes, which is almost entirely associated with excess body fat. Most people with type 2 diabetes get it as a by-product of overweight or obesity when the liver, skeletal muscle, and vital organs are infiltrated with fat. Fat in the liver interferes with the processing of insulin, making the liver relatively insulin resistant.

Why is insulin so important? What does it do in the body?
Let's use an analogy. Think of glucose, blood sugar, as someone trying to get through a door. In this case, the door is a receptor, and it leads into a cell. Insulin is the usher that opens that door. So essentially, insulin interacts with the receptor: It opens the door and lets glucose into the cell. With insulin resistance, the door is jammed, so the glucose can't get into the cells.

This means, for one thing, that your pancreas has to churn out twice as much insulin — because it's important for the glucose to get through the door: Your cells need it!

And insulin does other things, too: It's also a growth hormone that favors the deposition of calories into body fat.

Finally, there's stress in producing the extra insulin; at some point, your pancreas cannot meet demand. It just poops out.

What's really ominous about all this is that the degenerating cascade is self-propagating: Excess fat around the middle causes insulin resistance. Insulin resistance means the pancreas must make extra insulin. Extra insulin results in more fat deposition around the middle. That worsens insulin resistance and drives insulin levels higher still. This persists, and worsens, until the pancreas gives up in exhaustion. Diabetes ensues.

Can the pancreas ever recover?

In the short term, yes; however, diabetes tends to become permanent over time. If you've had type 2 diabetes for five years or longer, it becomes very unlikely that you will reverse it. Within those first few years, the pancreas can generally recover if you lose weight, and by doing so clear fat from the liver and improve insulin sensitivity there.

Why is there such an obesity epidemic in the United States?

Everything from what's grown to what's produced and sold, and how those foods encourage us to eat too much of the wrong things. (Stress and sleep deprivation also play roles.) The obesogenic environment makes it likely that most people will wind up overweight, and so we do. We are encouraged to get fat.

What about personal responsibility?

In this case, that's blaming the victim. Yes, we are all responsible for what we choose to put at the end of our fork. But (and we're going to soapbox it for a minute here) come ON! What about collective responsibility? The choices any one of us makes are always subordinate to the choices all of us have.

And we live in a food supply willfully designed by experts to maximize eating for the sake of corporate profits.

Personal responsibility is not a license for irresponsible public policy. Personal and collective responsibility are not mutually exclusive. Even if you have taught your child how to swim, you still count on rip tide warnings at the beach, right? You still expect the lifeguards to do their part, too — right?

When food choices are bad . . . it's hard to make the good choice. Somehow, we've wasted some time conflating the risk of chronic disease with value as a human being. That's changing: The recent body positivity movement embraces all sizes and pushes back against value judgments placed on people because of size. We live in a highly obesogenic culture, and no one makes it easy to eat the kind of optimal diet we're pushing so hard for here.

Still, obesity is problematic and diabetes has really gotten out of hand.
Yeah, it's horrible, but blaming overweight people and those with diabetes is all wrong. Bathroom scales and glucometers do not measure character or worth, and we have to unbundle disease and "personal responsibility." We have to be able to confront the health threat of obesity without blaming the victims of the condition for it. To give an analogy, when fossil fuels pollute the air, or create climate change, we encourage people to carpool or use public transportation. But it's equally important for society to turn to new forms of energy.

Suppose I need to lose weight. What can I do, eating-wise?
Foods that are high in fat and sugar are most likely to contribute to overeating. You know what these are. Junk food. Pizza. Ice cream. Soda. Donuts. You don't need us to tell you this.

So after I eliminate those foods, what should I eat instead to help me lose weight?

Eat lots of, yes, vegetables, fruits, whole grains, beans, lentils, nuts, and seeds. Most of these are filling and low in calories. Those that are high in calories, like nuts and seeds, are still nutrient rich and notably satiating in their unprocessed state. When you're thirsty, drink (plain) water. All this may not be easy, but it's pretty damned simple.

Minimally processed foods that are close to nature tend to be more satiating relative to calories, whether they are high in fat or high in carbohydrate. Even more specifically, foods that are high in fiber — so, beans, lentils, whole grains — are great for feeling full. We refer to oatmeal as a "stick-to-your-ribs food"; it really is, and that's because it has a high concentration of fiber.

But nuts are actually high in fat and calories. Shouldn't I avoid them if I want to lose weight?

Not at all. While they are high in fat, they are also associated with satiety, and they're highly nutritious. As with all food, of course, there are limits. And remember, what we do to a food matters a LOT. We are apt to eat plain, raw almonds until full; all good. We are apt to eat honey-roasted almonds — roasted, salted, sugared — until our arm gets tired from lifting them to our mouths! Not so good.

I know I'm not supposed to eat too many calories, but, well, what is a calorie?

A calorie is the energy required to raise the temperature of one cubic centimeter of water one degree Celsius at sea level; a kilocalorie is the energy required to do the same to a liter. When it comes to food, we use kilocalories as a measure of the energy that the food will provide. (Nutrition labels generally use the term "calories" to mean kilocalories.)

Do calories matter?
Of course they do. The amount of energy you put into your body determines many of your functions: Without enough, you die. With too much, the excess is stored as a reserve for future use, meaning you gain weight. If you restrict calorie intake, you will lose weight, almost without exception.

But are all calories the same?
They all contain the same amount of energy, but your body can respond to different *foods* differently, even if they contain the same number of calories. It's a mistake to think that eating too many calories is the *only* reason you can gain weight, and it's a mistake to think that since all calories are the same, energy-wise, all foods are the same. A calorie from black beans is not the same as a calorie from jelly beans.

Think about the analogy of gas in your car: The amount of gas you put in matters — your car won't run at all on empty. But of course it also matters whether that gas is clean, contaminated, or diluted with water. Likewise, you shouldn't think that all calories are the same — they're not.

How many calories should I really be eating?
You'll notice that the nutritional information given on food packaging is based on a 2,000-calorie diet. But it's a prototype — it isn't reflective of our diet, and probably not yours either. It provides a reference so that the units of measure for all foods and nutrients are relative to some standard. The number isn't arbitrary: In medicine, the time-honored anatomical prototype (for better or for worse) is the 70-kilogram man — that's a man weighing 154 pounds. (One of us pretty much is that guy — but there aren't many of us left these days!) The 2,000-calorie figure represents roughly the amount of food energy required to maintain the weight of that 154-pound

adult male, given a fairly average metabolic rate and fairly average activity level.

But while calories are relevant to right-sizing your diet, there are better approaches to losing weight than counting them. We've already said it: Aim for a balanced diet made up of wholesome, high-quality foods. Since plant-based foods high in fiber and healthy fats make you feel full and satisfied, you'll probably find yourself naturally consuming fewer calories.

Specific Diets

THE MEDITERRANEAN DIET

On page 16, you talk about the Mediterranean diet. I've heard that this is the best diet in the world.
The Mediterranean diet is not a diet in the fad sense, but a traditional way of living. It's not a strict set of rules like "eat grapefruit every morning and steak every night," but rather a lifestyle. Having said that, part of that way of living is eating in a way that works well for health. That makes the Mediterranean diet a candidate for "best." But it's not alone.

You're saying the Mediterranean diet is technically not even a diet?
Yep. In fact, that's a main distinction of the Mediterranean diet from other diets. We've sort of torqued it to be a "diet" because that's how we like to talk about our relationship to food. This is a mistake. What we really want is a dietary pattern we can live with for a lifetime, with other members of our family or our household or our community. A "diet" in the perverted modern sense is something we go on, and off; it is not a way of living.

This seems like an obvious question, but the Mediterranean diet is inspired by how people eat in the Mediterranean?

It's a general representation of key, common elements in the dietary pattern of people living around the Mediterranean. There are many variants: It can be quite different in North Africa than it is in the Middle East, or in Italy versus Spain.

Perhaps it's mostly that a traditional way of eating is healthful, or maybe that ways of eating that are not healthful don't tend to become lasting traditions!

But experts have highlighted the important common aspects and define the Mediterranean diet as one with lots of vegetables and fruits; whole grains; nuts and seeds; legumes; and an abundance of heart-healthy oils from olives especially, but also from avocado, those nuts and seeds, and, to a lesser extent, seafood.

Why do Mediterranean populations eat the way they do? To be healthy?

Since this is a cultural norm spanning many generations, it was not originally chosen for health — it's just how people wound up eating based on local agriculture and what thrived in the region. Perhaps it's mostly that a traditional way of eating is healthful, or maybe that ways of eating that are *not* healthful don't tend to become lasting traditions! (You think anyone is going to be eating the standard American diet (SAD) a hundred years from now? We don't.)

In any event, at this point, people around the Mediterranean eat this way because they always have — each generation learns what to eat, and what to like, from the generation before. They don't do it because they hope it causes weight loss in the short term, but for bigger, long-term reasons. Now, of course, we know all about the diet's sustainability, palatability, and how positively it affects people over the course of their life. So, now it's not just "how things are done here," it's also world-famous for being a very good idea.

Unfortunately, Americans have effectively exported junk and fast food, and that's changing the way people eat all over the world — including the Mediterranean.

The Mediterranean not-diet, or way of eating, or whatever you want to call it, sounds a lot like the general guidelines you've described elsewhere in this book. How is it different?
The thing that sets the Mediterranean diet apart from other plant-based diets is that it tends to be high in fat, particularly extra-virgin olive oil.

This goes back to the question of whether fat is bad for us.
Right, and as we'll discuss more starting on page 160, some fats are way better for you than others. But again, what we really want to do here is stop thinking about macronutrients as "good" or "bad." Balance is good, imbalance is bad, and foods that move us toward a healthy balance are good. Certain polyunsaturated fats and monounsaturated fats tend to move us toward that healthy balance, and extra-virgin olive oil is an excellent source of those. In addition, extra-virgin olive oil is rich in some potent antioxidants that may confer unique health benefits (see page 134 for more about olive oil).

Is the presence of fast food harming the overall health of the population? One Big Mac a week within an otherwise traditional Mediterranean diet I hope wouldn't do much harm.
Some of the latest reports from Mediterranean countries found that they are starting to succumb to obesity and chronic disease. It's not a one-cheeseburger-a-week situation. We're packaging their Mediterranean lifestyle as a diet, while exporting typical American junk worldwide.

So Americans export high blood pressure and fast food while traditional Mediterraneans export longevity and extra-virgin olive oil.
That sums it up pretty well. Yay, us.

Other than observational studies of populations, has any research been done on the Mediterranean diet?
There have been large randomized trials; mechanistic studies of extra-virgin olive oil and its antioxidant compounds; and much more. One ongoing randomized trial is called PRED-IMED. An earlier randomized trial called the Lyon Diet Heart Study showed that the Mediterranean diet could slash rates of heart attack in Europeans at high risk compared to standard northern European diets. There's also a researcher from Athens, Antonia Trichopoulou, who is the "godmother" of the modern understanding of the Mediterranean diet. She's studied it the longest and most intently, and some years ago wrote a paper entitled "Anatomy of Health Effects of Mediterranean Diet."

And?
She identified key components including, as we've said, an abundance of beans and legumes as principle protein sources, daily servings of fruits and vegetables, whole grains, and a significant portion of fat from olive oil with additional fat from avocado, nuts, and seeds. She said there are small amounts of meat, limited amounts of dairy, and some wine. Depending on the place, fish and seafood is sometimes in there.

So if I drizzle olive oil over my fries, that's Mediterranean?
Not at all; what defines a diet is the overall pattern of foods and their proportions, not just adding one ingredient or taking something away. There is, in fact, a scientific scoring sys-

tem developed to assess adherence to the true character of the Medi diet. (If you want to find it, look for "The Reliability of the Mediterranean Diet Quality Index (KIDMED) Question- naire.") But you need not go so far to know whether your diet is overall Mediterranean-like or just business as usual, doused in olive oil. They are not the same!

Still, what's the scoring system?

If your diet includes food that's typical of the Mediterranean diet, you get more points. If it excludes component foods or includes unhealthy foods, you get negative points. So, for ex- ample, you get more points for eating bulgur and garbanzo beans than for salmon. But you would get negative points for eating French fries. Even if you drizzled olive oil on them.

This seems helpful!

A point system is like training wheels on a bicycle: It helps you learn. And that makes sense when you consider that the most famous and prominent weight loss approach in the US has been Weight Watchers. It also makes sense when you consider what the default dietary pattern is in the US these days.

Which is?

The "see food" diet: See food and eat it. Whatever, whenever, however, and however much that happens to be. If you're try- ing to resist that and adopt the dietary patterns from a differ- ent culture, like the Mediterranean diet, defining it a bit more precisely is important, at least at first.

Once you know you can do this, then you can live more freely. Take the training wheels off. But the metrics are a good start for someone who says, "I want to go Mediterranean, get it right, and see the benefits."

VEGANISM AND PART-TIME VEGANISM

Should I go vegan?

To start with, veganism, just like the Mediterranean lifestyle, has been repackaged for popular consumption as a "diet." But veganism, and part-time veganism — a less dramatic, more mainstream, and equally sound variation, at least from a health point of view — reveal that all good diets have this in common: They rely heavily on calories from minimally processed plants. Part-time veganism is a variation on the Mediterranean diet, and vice versa — and so, too, are the many other variants on the theme of optimal eating for health.

> It's not strict veganism that's important, but the overwhelming predominance of plants in a diet.

Who lives a vegan lifestyle, other than modern-day adapters?

There are large populations around the world that are either vegan or nearly so. There are whole portions of Indian populations that are vegetarian, if not predominantly vegan, and Seventh-day Adventists are either vegetarian or vegan much of the time. And of course there are people around the world who are vegan by necessity; they simply have access to no or very few animal products.

It's not strict veganism that's important, but the overwhelming predominance of plants in a diet. Being 90 percent vegan, from a personal heath perspective, is probably just as good, perhaps even better. Veganism, per se, is generally more about advocacy and activism on behalf of animal rights than health. And these days, many people are eating vegan because of the environmental footprint of animal foods, which is astronomically higher than that of plants.

Yeah. There are definitely stronger motivations for people to be vegan than just their waist size.

Most people who practice veganism with real fidelity are thinking beyond the bounds of their own skin. It's not just "This is good for me." While their diet may very well be good for them, committed vegans are also thinking about two other important issues — ethics and environment.

I mean, there is no question that a diet that doesn't involve killing or harming animals is kinder and gentler to creatures than any other diet. What about the environmental impact?

Animal foods are much less efficient to produce than plant foods. As you move up the food chain, energy is lost at each stage. So the environmental impact of food production would be reduced if people in developed countries ate a lot less meat and dairy.

What isn't as clear, in terms of the ideal balance between low environmental impact, nutritional quality, and feeding the world's billions of people, is the ideal mix of plant and animal foods. In other words, we know what direction we should be moving in as a population. We just don't know exactly what the future will look like.

If energy is lost at each stage, does that make plants the highest-energy foods?

Well, the most energy efficient, not the highest in energy. Energy comes from the sun, and plants convert sunlight to stored energy; not perfectly, but with high efficiency. Animals that eat plants convert that energy in those plants to their own energy, again, imperfectly. And animals that eat animals repeat the process. So energy is lost at each step.

The most efficient way to access the energy of sunlight is to eat the plants that captured it, rather than eating animals that ate the plants. When we feed animals to create the meals that we eat, we give up a good amount of energy that was intrinsic to those plants in the first place.

While there are all sorts of debates about optimal food production systems and sustainability — whether animal products should be involved, and if so how much of them — it's clear that shifting toward more plant-based diets has significant benefits for the environment.

So can we confidently say that the best diet for the environment is veganism?

For you as an individual, yes: The world eats way too much meat and dairy than is good for animals and the planet, and each time one of us stops doing so, we help move that total intake in the right direction. But it's harder to say that veganism by all humans would be the most efficient use of resources. There's debate about that, and no one has a clear, decisive answer.

But the direction we should all be moving in — toward more plant foods, fewer animal foods — is clear and noncontroversial. And because modern diets in the West tend to be animal-food predominant, a shift in the direction of vegan-like dietary habits would be a really good thing for personal health, too. For most of us, that would likely mean a mostly or totally whole-food, plant-based diet — veganism, or flexitarianism, reduceatarianism, or simply plant-heavy omnivorism — whatever you want to call it.

To be clear: Eating a whole-food, plant-based diet (or mostly so, with animal products seen as a treat or a garnish rather than a staple) is the most realistic, beneficial, fair, sustainable diet for most people and the planet. Not coincidentally, it's not that different from the Mediterranean diet.

But "vegan" doesn't necessarily mean "healthy," right?
"Vegan" is simply a label that could be applied to diets that vary dramatically in practice. A diet associated with famine, limited to cornmeal and little else, might be vegan. A diet of Coca-Cola and cotton candy might be vegan, too.

But a vegan diet can be *very* high quality, and some argue it is the best diet for human health. That's controversial, but certainly an optimal vegan diet is among the contenders. Vegan not by choice may mean hunger and an incomplete diet; but vegan by choice can be complete and balanced, or otherwise. That's an important distinction to make; it's a privilege to be able to experiment with different diets. And when veganism is the choice, it is often about a trifecta of interests — the environment, the ethical treatment of animals, and health — rather than just one.

Doesn't being vegan put you at risk for protein deficiency?
No, and no, and no: That is a myth. A balanced vegan diet provides all the essential amino acids and plenty of protein, even for the bodies of world-class athletes. Protein deficiency occurs when diets are inadequate, not because they are vegan (see page 151 for more on protein).

Does it matter how much fat or sugar you eat if you're vegan?
Veganism, per se, says nothing about sugar or fat. And, as we said, you can definitely do veganism badly; just because something is vegan doesn't mean it's healthy, whereas if you follow Mediterranean rules, you're guaranteed to be eating well.

That's because veganism is defined by what is excluded — animal products — whereas the Mediterranean diet is defined by the foods included. The "vegan" answer to that, and the increasingly popular term, is "whole-food, plant-based." That does emphasize what is included: whole plant foods, notably vegetables, fruits, legumes, nuts, seeds, and grains (and no longer allows for the Coca-Cola/cotton candy diet). Note, also, that what is emphasized here is similar to what is emphasized in the Mediterranean diet. What sets the vegan rendition apart is that it is whole-food, plant-*exclusive,* whereas all other truly healthful diets are whole-food, plant-*predominant.*

Can I really get all my nutrients and protein from eating just plant foods?

The fact that you may not get all the nutrients you need from a dietary pattern is in no way unique to the vegan diet experience. Most older Americans get less zinc than recommended. Almost everyone who works indoors, and wears clothes, gets less than the ideal amount of vitamin D. That said, pure or strict veganism does tend to result in low levels of long-chain omega-3 fats (from so-called fish oils, though there are also plant sources) and vitamin B12. With regard to micronutrients, then, part-time veganism may be even better, because some nutrients are more reliably found in animal products. (Although there are, by the way, also plant-food sources of vitamin B12, like nutritional yeast.) But even if a vegan needs to supplement with B12, so what? Most American diets are deficient in certain nutrients. While you're at it, you probably ought to supplement omega-3s and vitamin D, too.

Should I be a low-fat or a high-fat vegan?

It doesn't matter, assuming the fats you get are from wholesome sources, and in balance — there's a tendency toward too much omega-6 in the diet (see page 131). But there are two schools of thought about the vegan diet: One emphasizes that as long as you are eating mostly whole plant foods, you're good. This would allow for plenty of avocados, nuts, seeds, and added oils, such as extra-virgin olive oil. We fall in this school.

The other school . . . ?
Is very wary of fat. They are a quite ardent camp, and believe that any added oil in the diet, including olive oil, is toxic. (We talk more about that on page 137.) Our advice is to recognize that advocacy for the low-fat version of a whole-food, plant-based diet is coming from a blend of epidemiology (the science and study of health in human populations) and ideology. If you want to base your own decisions just on epidemiology, you really do have choices. In our view, it's not low-fat versus high-fat that matters, but the fact that a diet is wholesome, balanced, and plant-predominant. Both Mediterranean and whole-food, plant-based meet that qualification.

Where did the evidence for low-fat vegan come from?
Most famously from Dr. Dean Ornish, a physician, lifestyle medicine expert (and friend of both of us), who did a study very early in his career that put people with advanced coronary artery disease on an ultra-low-fat vegan diet. After a span of months, repeated coronary angiography showed that the plaque had regressed. In other words, an ultra-low-fat diet reversed heart disease.

That sounds legitimate.
It does, and it is. And a later study of the same group showed that the intervention did, indeed, slash rates of heart attack over time. And observational research by Dr. Caldwell Esselstyn at the Cleveland Clinic showed much the same.

But the only direct evidence that a low-fat vegan diet causes plaque in the arteries to shrink and go away is limited to these examples, which doesn't prove that other diets can't do the same. Other diets simply haven't been studied in just this way, and that leads to an important, general rule: Absence of evidence (i.e., something hasn't been studied) is NOT the same as evidence of absence (i.e., it's been studied and shown not to work).

Huh?

Would a solid Mediterranean diet also cause plaque to shrink? Probably. But the big studies of the Mediterranean diet did not do serial angiography. They did, however, look at heart attacks and death rate, and both are lower with the Mediterranean diet. From our point of view, following populations over years to see who did and didn't get heart attacks, who did and didn't have strokes, who did and didn't die prematurely is more important than establishing what coronary arteries look like.

We care about plaque in our coronary arteries because it predicts the risk of cardiovascular disease. So if we know the answer to the definitive question (does a diet prevent heart attacks?) the surrogate measure (does a diet reduce plaque?) no longer matters. This is why, in our view, low-fat versus high-fat is just not important once you get the dietary pattern right. Both approaches have been shown to add years to lives.

So that's what matters, then?

Isn't it? If our blood pressure, or cholesterol, or arterial calcification, or pick whatever "biomarker" you like, did not tell us anything about our likely longevity, or vitality, why would we care about them? We care about measures of health that indicate what the length and quality of our lives and health are apt to be. The reason we *do* care if our coronary arteries don't look too good on a CAT scan, MRI, or angiogram is because those measures predict if heart disease is likely to harm us or cause premature death.

Let's say I'm a vegan mainly for health reasons. Would I benefit from adding fish?

We don't have enough evidence to definitively answer that. We know that in general eating fish is "good for us." But inevitably that comparison is between modern, mainstream diets that do or don't include fish. In that context, fish is almost always displacing meat: That's definitely good. The study hasn't been done that compares a balanced, whole-food vegan diet versus one with fish. Putting an optimal vegan diet and an optimal pescatarian diet head-to-head and figuring out where the trade-offs are between fish and lentils, or fish and beans, would be interesting. For now what we know for sure is that eating fish is good for people who use it to replace pepperoni, salami, hot dogs, and red meat in general.

What does a part-time vegan diet look like?
Like a Mediterranean, flexitarian, or vegetarian diet. It allows for some flexibility in place of very strict rules, which can make it easier to follow a plant-predominant diet.

What does it mean to be a flexitarian?
The flexitarian diet, when done well, is one that emphasizes a wide variety of highly nutritious plant foods but allows for small amounts of dairy, eggs, meat, poultry, fish, and seafood. It's another way of describing part-time veganism, really (and vice versa).

And what does it mean to be vegetarian?
Well, different things to different people. In general, it means no meat — a mostly vegan diet plus eggs and dairy (often called "lacto-ovo vegetarianism"). Some people include dairy but not eggs, or vice versa. Allow a little bit of fish, and you'd be a "pescatarian"; do the same with poultry or meat, and you've crossed into flexitarianism. But these are all variants on the theme of plant-predominant diets, and often blended in the real world.

The line between plant-based diets is actually quite blurry.
It's a blurry line because they all are healthy options! If you
want to eat a whole-food, plant-predominant diet, you don't
need to shift very much to find yourself in the context of a
Mediterranean diet, a DASH diet, flexitarianism, pescetarian-
ism, or part-time veganism, however you define it.

INTERMITTENT FASTING

I've heard a lot about intermittent fasting. How does it work?
The most popular version is to eat on five days and fast for two,
so your net intake of calories for the week is lower. And, for
most people, that's the goal of intermittent fasting.

Is it safe to go 24 hours without eating something?
It depends. For healthy people without severe metabolic de-
rangement, the answer is almost always yes. In fact, we are
probably well adapted to intermittent fasting if we look to our
Stone Age ancestors.

Humans aren't genetically designed to eat three meals a day?
Nope. In the Stone Age, people would eat what they could find
when they could find it. Sometimes they couldn't find much
of anything. We have good reason to believe that intermittent
feast (and famine) was native to the human condition. We are
actually quite well adapted to periods of fasting.

**The evolutionary argument seems like a moot point. Haven't
our lifestyles changed significantly since the Stone Age?**
Modern changes don't make evolutionary arguments moot.
Circumstances can change quickly; adaptations baked into
our DNA because of survival advantages generally happen
more slowly. Our genes still carry the imprint of many Stone
Age adaptations.

But our modern circumstances do matter, don't they?
Yes, of course. Our Stone Age ancestors didn't eat whenever they wanted to — they fasted intermittently whether they wanted to or not — and we are probably adapted for that. But they did not have diabetes or congestive heart failure, either, and such conditions are now common. A majority of adults in the US have one or more chronic diseases and take medications for them; that changes the equation.

How does it change my decision about fasting?
Fasting is definitely not safe for everyone, especially for those on medication. If you're using insulin to keep your blood sugar in balance, for instance, and you radically shift the pattern of your diet day to day, you are going to have difficulty keeping your blood glucose in a healthy range. If you're on blood pressure medication, fasting could make your blood pressure dangerously low.

If the only rule is that I can't eat on certain days or at certain times, does that mean I can indulge a bit more because I'm starving myself the rest of the time?
No, what you eat remains important. Intermittent fasting doesn't address the *quality* of your diet; it's primarily a weight loss scheme. But (and we'll get into this in detail later) both calories and nutrition count, and weight is a factor in health. One of the downfalls of most intermittent fasting schemes is that they stipulate virtually nothing about what you're eating — they only emphasize that sometimes you won't be eating anything. Our concern is that this may suggest that as long as you don't eat anything a couple of days a week, you can eat anything you want the other five. Nothing could be further from the truth: If you binge on your higher-calorie days, or in the meals you're not skipping, all bets are off.

If you are going to use intermittent fasting as a means of

controlling weight, make sure that on the days that you do eat, you pay attention to what you are eating.

I have to admit that when I skip a meal, I eat much more later; I can't control myself.
Then the intermittent fasting diet is probably not the best choice for you. It works for some people, and doesn't for others.

If I get faint or grumpy from not eating, are there other ways to restrict my calorie intake?
Of course. It's called portion control.

Does one work better than the other? I've heard that fasting revs up your metabolism in a way portion control doesn't.
They work almost identically. Proponents of intermittent fasting argue that turning the metabolic furnace on and off, as it were, uses up some extra calories, and studies suggest they could be right. But those studies also show that the effect is almost trivial when compared to controlling calorie intake overall. So really, fasting is just one tactic for calorie or portion control; claims of "revving" your metabolism are certainly more sales pitch than science. Overwhelmingly, the weight loss effect of intermittent fasting is from reducing calories in, not an increase in calories out (i.e., burned by your metabolism).

Proponents of intermittent fasting claim it fosters enhanced mental clarity, too, but we find no science to say this is a consistent effect. Some people use fasting days as a "cleanse" and combine it with meditation; in such cases, enhanced focus is likely a result of . . . well, focusing.

Lastly, we've encountered claims that intermittent fasting leads to enhanced longevity. This is likely a trespass into the realm of caloric restriction, where there is evidence in other species of a longevity benefit. So, if you fast enough to stay

30 percent or so below your conventional calorie require-
ments for long enough, there might be a longevity benefit. Or
not. Let's be blunt about it: No one has done the hundred-
year-long study to know for sure!

If it's a tactic that helps you control the total amount you
eat on average over, say, a given week (and then month, and
then year), then it's advantageous for you. But that's just not
true for everyone, and any tiny "metabolic" benefits are easily
overwhelmed by overindulging.

So, no: There's no metabolic magic involved in fasting.
Fasting helps some people, but not others, just like all other
nonmagical approaches.

Then which should I choose?
It comes down to which method you prefer. Most people prefer a degree of dietary restraint to nearly starving themselves
a couple of days a week, and that can work just as well. If you
prefer to be somewhat more indulgent five of the seven days
and compensate with two days of fasting, that can work, too.

**Intermittent fasting seems like it could be a real pain for
everyone else around me . . .**
Agreed. If you have a family and you fast but they don't, things
can get awkward. Do you sit at the dinner table and . . . twid-
dle your thumbs? Take charge of scintillating conversation?
That could work, but obviously there are potential issues. And
there's more to consider: Can you be productive on your fasting days? What's your level of mental acuity? Your employer
may not be thrilled if you are slumping over in a famished tor-
por. So maybe you want to fast on the weekend. But isn't that
when you want to indulge a bit?

Those are vital considerations that are often overlooked regarding all odd diets. In unity there is strength, right? When
we talk about dieting, it's usually a go-it-alone thing (that's

why we contrasted "live it" and "diet" earlier; we tend to diet alone, but could live it together). One of the many reasons dieting has failed us is that we try to go at it alone. That's another reason why a moderate, traditional dietary pattern, one that's not "weird," is a better choice.

But would you recommend intermittent fasting if my goal is to lose weight?

When all is said and done, if it works for you, yes. Intermittent fasting is a currently popular "let's go on a diet" sort of thing. Unlike many fad diets, there are few reasons against it, so if it suits your taste and lifestyle, go for it. But it's not a short-term fix.

PALEO DIETS

Tell me about the Paleo diet. That's the "caveman" diet, right?

Paleo claims to mimic the diet of our foraging, hunter-gatherer Paleolithic ancestors. ("Paleolithic" refers to the "old" Stone Age, the long period from roughly 2 million years ago to 10,000 years ago, during which various human species used rough stone tools. This was followed by the Neolithic, or "new" Stone Age, during which agriculture was introduced.)

What did cavemen and cavewomen eat, exactly?

People can pretend to know, but no one really does. We have a very hard time accurately measuring what people ate yesterday, so let's allow for error bars around what our great-great-great-great-(etc.)-grandparents ate 127,000 years ago.

And because of those error bars, there are ongoing debates among card-carrying paleoanthropologists. Some contend that our forebears mostly hunted; others contend that they mostly gathered; others suggest that insects were a major dietary contributor; and others claim that food from the sea was of high importance. The field is advancing all the time with studies of archeological finds, including analysis of dental wear patterns (how teeth are worn down), skeletal remains, campsite remains, and even coprolites (fossilized fecal remains).

So, we start with: Who knows, exactly? But what we do know is they foraged and hunted; they didn't farm. So: meat, fish, eggs, nuts and seeds, vegetables, fruits. Whatever they could find or kill. Obviously, these were all "natural" foods. They ate wild plants, and wild animals that had eaten wild plants. We can be quite confident about many things they didn't eat: pepperoni and bacon; donuts and toaster pastries; cheese and soda. So a Paleo diet eliminates processed foods, including added sugar, refined grains, legumes, and dairy; there's debate about whole grains, and many Paleo proponents exclude them. But some of those expert paleoanthropologists have found evidence of wild grains in our diets going back 150,000 years or more.

Are there foods in the modern Paleo diet that our ancestors may not have been eating?

Some selected fats and oils, like coconut oil and butter, and some processed sweeteners, like coconut sugar. But today's meat is also wildly different, no pun intended, from pre-agricultural meat; today's animals are raised in captivity and eat differently, which changes the quality of their meat.

Also, to put it quite bluntly: Everything our actual Stone Age ancestors actually ate is, well, extinct, mostly thanks to us. We don't eat any of the wild plants they ate; we eat domesticated plants, sometimes closely related, sometimes not. We don't eat wild animals (for the most part), and when we do, they're not the same wild animals. (Seen a mammoth any time recently?) In fact, it was the domestication of plants and animals that ended the Stone Age and ushered in the age of agriculture.

The Paleo diet doesn't allow for dairy, legumes, and grains because it eliminates food that didn't exist before agriculture?

That's the idea. Of course, you can brand anything you like "Paleo," so some proponents include butter, which is obviously a stretch.

And yet Paleo allows for coffee and a dollop of ghee.

Yup. Despite the fact that there was neither coffee, nor ghee, in the Stone Age. Call it poetic license. Or . . . nonsense.

What else sets Paleo apart? Foraging berries for breakfast and hunting for dinner?

That's a pretty unrealistic expectation for most, but the underlying ideology encourages regular exercise — usually shorter, intense workouts — and a mindfulness practice, like yoga. But as is often the case in a sound-bite, clickbait culture, a lot of the inconvenient caveats and provisos are dropped on the way to "I can eat as much bacon as I want? Sign me up!"

It's not hard to do better than the typical modern American diet, and any reasonable interpretation of a Paleo diet almost certainly does.

The experts suggest that our Paleo ancestors walked miles every day, and consumed about 100 grams of fiber daily from a variety of wild plant foods. (We currently eat about 15 grams a day.) There are very few modern Paleo practitioners aiming for that, which is too bad.

Our point is that the Paleo diet has really been much reinvented for modern consumption, and convenience. It's the very rare practitioner who is attempting to emulate a true Paleo lifestyle, inconveniences and all, with any fidelity.

Are there health benefits to eating like our ancestors did forty thousand years ago?

The claimed benefits are weight loss and control, better digestion, and lower environmental impact. There is also some evidence suggesting that the Paleo diet is anti-inflammatory, prevents cancer, and lowers cardiovascular risk. This is based on short-term studies that use various biomarkers like blood pressure, blood lipids, or weight; there are no randomized intervention trials on this topic lasting the years or decades it would require to know this for sure, and no parts of the world where loads of people have been eating "Paleo" for generations. But it's not hard to do better than the typical modern American diet, and any reasonable interpretation of a Paleo diet almost certainly does.

That all sounds, um, pretty awesome.

It does, but there's more going on here. Remember that how diet affects health depends entirely on what the diet is replacing. The Paleo diet is a vast improvement over the typical American diet; if you replace Twinkies with vegetables or donuts with eggs, you are obviously going to see a health benefit. But while many advocates claim that eating Paleo is the best diet, there is not a lot of evidence to support that. (Nor is there evidence that it causes harm.)

But Paleo is really different from a whole-food, mostly plant-based diet, right?

Actually, a true Paleo diet and a vegan or mostly vegan diet are more alike than different, especially placed against the backdrop of the typical American diet; they both emphasize real foods in as-close-to-natural-as-possible form. The difference is that Paleo diets include meat, ideally from wild animals, but leave beans and grains behind. But a sound Paleo diet, one that includes lots of fruits and vegetables and not a ton of meat, has a lot in common with a plant-based diet. This is the surprising take-away: While they may seem polar opposites, a really good practice of the Paleo diet and a really good practice of veganism are far more like one another than either is anything like the typical modern diet of glow-in-the-dark "food-like stuff."

Are there any fundamental problems with the Paleo diet?
Yes, three. The first is what we call "the pastrami problem," and the second, "the population problem." The first is that many people love the Paleo diet as an excuse to eat, well, pastrami. And any other meat they like, and lots of it. There's this false idea of "forget the bread and it's suddenly Paleo."

Obviously there was no Paleolithic pastrami.
Nope. Our ancestors only ate the wild animals they hunted themselves, which were nothing like the fattened, domesticated animals used in the production of processed meat.

So bacon and eggs fried in butter isn't really a staple Paleo breakfast?
No. And bacon is totally different than antelope. If you want a Paleo diet, our contention is that you have to practice it with true loyalty if you want to see any health benefits at all, which means occasional meat only, and as close to wild as possible, with big-time reliance on foods that were once gathered: fruits, vegetables, nuts, seeds, and eggs.

What do you mean by "the population problem"?
This has far more grave consequences than the pastrami problem. There are more than 7 billion humans (nearly 8) on the

planet. Back in the days of hunter-gatherers, each tribe of 100 people needed about 32 square kilometers of land. For 7 billion of us, that would be 15 times the entire land surface of Earth. So with a real Paleo diet, even if we used every inch of land — and found food there — only 6.7 percent of us would avoid starvation. So it's not really about whether an optimal Paleo diet is better or worse for human health than an optimal vegan diet. The relevant question is: What diet is really a viable option for a growing population on a shrinking planet?

A Paleo diet won't work for everyone.
Right. It's true that Paleo diet enthusiasts could live alongside vegetarians and vegans. But keep in mind that a true Paleo diet is still plant-based with a moderate amount of free-ranging animals. A real Paleo diet doesn't allow for factory-farmed meat, the abuse and confinement of animals, or animals eating anything other than their native food.

What's the third problem?
We just don't know what this diet does for lifelong vitality and longevity in comparison to other good diets. A diet of a variety of wild plant food, and the meat of wild animals would probably be very good for us. But it's not proven one way or another.

Anything else to say to Paleo enthusiasts or potential converts?
Paleo enthusiasts can legitimately eat their local beef and wild-caught fish, and might potentially derive real health benefit from doing so, allowing for the above.

But Paleo is not a sustainable diet for most humans. We don't live in an age of isolated, scattered human tribes; we live amid a modern population of 7 billion, and it's growing fast. A mostly plant-based diet is the common interest of any of us who want to stick around.

THE DASH DIET

What does the acronym DASH stand for?
Dietary Approaches to Stop Hypertension.

Sounds like it has a very particular medical goal in mind.
It was developed in the 1980s by the National Heart, Lung, and Blood Institute at the NIH. The original hypothesis being tested was whether a dietary pattern could reduce blood pressure as effectively as drugs. The answer was a resounding "yes," and it turned out to be a generally sound diet in the bargain.

What dietary change could reduce blood pressure as effectively as drugs do?
DASH emphasizes whole plant foods, notably whole grains, low- or nonfat dairy, and reduced sodium intake. The more sodium consumed, the higher blood pressure tends to be.

So a main principle of DASH is to restrict sodium intake?
Yes, although the diet was actually studied with and without specific levels of sodium restriction, and worked both ways. But the combination of a sensible, balanced, plant-predominant dietary pattern with low-fat dairy AND sodium restriction worked better than either alone.

But salt makes food taste so good!
The DASH investigators had that concern, too. They decided that if the DASH diet needed to minimize salt, it could embrace herbs and spices for flavor. Also — fun fact — about 80 percent of the salt in our diets is processed into foods, not shaken on by us. So, the best way to reduce sodium does not involve banishing flavor from the table; it simply involves eating fewer highly processed foods.

Does what I eat (like fruits and vegetables) versus what I restrict (like sodium and saturated fat) have an equal effect on lowering blood pressure? Or does one matter more than the other?

The combination of healthy eating and restricting "bad" foods is the most effective way to go. Studies show that just restricting salt did help lower blood pressure, but they also show that the DASH diet lowered blood pressure even without restricting salt. As noted, combining the two was far more potent than either alone.

Easy enough. Are there any foods that the diet encourages that help reduce blood pressure?

Vegetables, fruits, and whole grains; maybe to a slightly lesser extent beans, lentils, and seeds; some fish and seafood, some poultry, and an allowance for some lean beef; and then low-fat and nonfat dairy.

Dairy, huh? This is the first diet we've talked about that includes dairy.

That's not quite true; the Mediterranean diet certainly allows for dairy, it just doesn't emphasize it. DASH grew out of the dietary guidelines in the US, where dairy has long been expressly recommended (whether for reasons of science or of effective lobbying is a contentious area).

But either way, low- or nonfat dairy is the actual recommendation in DASH in order to minimize saturated fat. Other than that caveat, dairy provides a lot of the vitamins and minerals found to lower blood pressure.

The best way to think about the DASH diet and contextualize it here is that it is a massively cleaned-up version of the typical American diet.

It allows for many of the foods that Americans eat —

some meat, poultry, egg, dairy — but it cuts way back on the sodium, ultraprocessed foods, and refined grains. It encourages routine consumption of vegetables and fruits every day, but it still allows for all the rest, albeit in moderate quantities.

Those sound like very realistic goals.
Exactly. DASH is not a radical departure from the foods that the typical American is eating. If you want to improve your health and weight but don't want to do a whole lot of heavy lifting, the DASH diet is a good option for you. DASH has generally come in right near the top in *U.S. News & World Report*'s annual best diet rankings, and we've just summed up some of the reasons why.

In comparison to something like the Mediterranean diet or a variation of the vegan diet, DASH sounds doable. But are the health benefits as profound as those of other diets?
Hard to say, for want of direct comparisons. DASH can rival some of the other dietary patterns. However, the magnitude of health benefit is likely to relate to the magnitude of dietary overhaul. The benefits for health and lifespan from DASH may be more limited precisely because you are making less radical shifts.

DASH kinda sounds like an American variation on the Mediterranean diet. What distinguishes the two?
Fat, mostly, along with the DASH focus on sodium. The typical American diet gets 34 to 35 percent of calories from fat. DASH would be down close to 30 to 32 percent, while the traditional Mediterranean diet would be up around 40 percent. While the Mediterranean diet allows for some dairy, there is no emphasis on low-fat dairy, as there is in DASH.

What's the conclusion? Does the DASH diet work as effectively as drugs?

For blood pressure, yes. The combination of the dietary pattern and sodium restriction lowers blood pressure as effectively as first-line drugs.

So it's food as medicine?

The DASH diet is just one glowing example of that. The nutrition community started advocating for DASH and asking other questions about the DASH dietary pattern: Is it effective for weight loss? Is it effective for other cardiac risk measures?

Well, is it effective for weight loss and heart health?

Yes. The DASH diet is a lot better for weight and health than the typical American diet. Since that's where the typical American is coming from, a shift to DASH is almost always a shift to the better.

ANTI-INFLAMMATORY DIETS

What's the hype about anti-inflammatory diets? I see "anti-inflammatory" as a benefit on a lot of foods.

We explain inflammation in more detail starting on page 168, but in short: Certain foods encourage inflammation. An anti-inflammatory diet avoids these foods. The body makes compounds called prostaglandins that can be either pro-inflammatory or anti-inflammatory. Both saturated fat and omega-6 unsaturated fat are building blocks for pro-inflammatory prostaglandins. But as we discuss on page 131, too much omega-6 — which is abundant in ultraprocessed foods — is not a good thing; it can contribute to excess inflammatory response.

So this is yet another reason to steer clear of ultraprocessed foods?

For several reasons: Many ultraprocessed foods contain omega-6–heavy oils as well as highly refined carbs, like white flour and added sugars. Starch and sugar don't directly flow into the production of inflammatory prostaglandins, but they drive up insulin. And insulin — a growth and stimulating hormone — enhances the activity of the immune system in a way that increases inflammation. Finally, diets high in ultra-processed foods foster obesity, and excess fat cells contribute more inflammatory compounds to the bloodstream.

Are there certain fats that aren't inflammatory?

Yes. Some fats are even anti-inflammatory. Omega-3 fat is most associated with anti-inflammation. Monounsaturated fat, the kind that predominates in olive oil, is less distinctly anti-inflammatory, but does not increase inflammation, either.

What other foods should I be eating to keep my immune system in balance?

The very foods you should eat to balance health overall, and by now you know the drill: vegetables and fruits, whole grains, le-gumes, nuts and seeds. For optimal health, these are the foods to increase. And sure enough, all of these are associated with dialing down inflammation.

Is an anti-inflammatory diet something most people should worry about?

Well . . . worrying is inflammatory! So, let's say it's something worth being conscious of.

Insulin resistance (prediabetes) is widespread in the United States, and remember that insulin is pro-inflammatory. Fixing this with diet and lifestyle is pretty much a universal priority. But fixing it with diet is easy, and requires no special effort, as long as you're fixing your diet anyway: You just need to move toward an optimal diet for general health. You can think about it in terms of anti-inflammatory, or you can think about it in terms of "healthy." You'll be making all the same adjustments either way.

Are there other contributing factors to inflammation besides food?

Absolutely. Psychological stress is inflammatory; it is associated with stimulation of the adrenal gland. Sleep deprivation is associated with disturbances in hormonal balances that increase inflammatory responses. So although the narrative has become "inflammation is bad and you want an anti-inflammatory diet," the underlying reality, again, is that *imbalance* is bad. The prevailing imbalance is what causes excess inflammatory responses, and can be addressed, fortunately, by the same good kinds of eating patterns we're talking about here.

THE LOW-FODMAP DIET

What's the low-FODMAP diet? What the heck does FODMAP even stand for?

FODMAP is an acronym for fermented oligosaccharides, disaccharides, monosaccharides, and polyols (or sugar alcohols). Now you understand it perfectly, right?

Sure. Translation?

These are all names for sugars and sugar derivatives. Some of them occur naturally — milk, honey, and fruit all have sugars — some are formulated. (Almost every sweetener that ends in "-ol" is ultraprocessed.)

The FODMAP concept is that the presence of these ingredients in foods will cause GI distress and other symptoms in susceptible people. Our body absorbs sugar alcohols — partly fermented sugars that can occur naturally or be industrially produced — poorly, and some of us are especially intolerant of them. Sugar alcohols tend to stay in the GI tract, where they can prove irritating.

What does the low-FODMAP diet eliminate, and why?

Anyone wanting to try the low-FODMAP diet will need to get detailed, written guidance, because there are a lot of particulars. But at a high level, this diet cuts out grains, onions, and garlic to avoid oligosaccharides; it cuts out dairy to avoid lactose (a disaccharide); it cuts out many fruits, honey, and added sugars to avoid monosaccharides such as fructose; and it cuts out some fruits such as blackberries that contain naturally occurring polyols (or sugar alcohols), along with gums and candies and mints that contain related, manufactured sweeteners like xylitol and sorbitol.

The main focus is on relieving symptoms of food intolerance; this is a kind of elimination diet. But just as gluten avoidance became thought of as a way to lose weight, or be healthy in general, the same sort of thing has happened with FOD-MAP: It's a diet that's become thought of as a cure-all.

If we absorb all these sweeteners poorly, wouldn't everyone be better off on the low-FODMAP diet?

Some of the low-FODMAP restrictions — such as avoiding artificially derived sugar alcohols — could be a good idea for all of us. But no, it's not a good idea to avoid grains and onions, along with many fruits and vegetables, if you don't have a good reason.

Who might benefit from a low-FODMAP diet?

Everyone should be avoiding ultraprocessed foods, but as for the rest, if you're having GI ailments, fatigue, brain fog, or any vague symptoms that you can't attribute to anything in particular, you could think, "Did these symptoms show up when I was eating a member of the FODMAP group?" The problem is that the group is so big, that's a hard question to answer. Depending on the severity and frequency of your symptoms, it might be worth exploring.

One notable example: Lactose intolerance is a FODMAP intolerance. Lactose is a disaccharide; two individual sugar molecules, glucose and galactose, are fused together to create lactose. People who don't have the enzyme to cleave them apart are lactose intolerant, and that's quite common. But lactose intolerance is usually dealt with by avoiding, simply, lactose. If that doesn't work for you, you might then try avoiding other FODMAPs.

The other entries on the FODMAP list can't be traced to a specific enzyme, and are more like gluten; some people just don't digest them well, or have an allergy-like response to them. Generally, the only way to know for sure is to remove the foods, see if symptoms go away, then add them back and see if the symptoms return.

Such "elimination diet" testing really is more robust for food intolerances and allergies than any other approach, and if you're suffering, it's worth a shot. If your answers are, "Yes, symptoms go when I stop, and come back when I resume," then you are among those who could benefit from avoiding FODMAPs.

As noted, FODMAPs reverberate widely through the food supply; the more narrowly you can restrict your diet to avoid only what ails you, the better.

Just as all diets have skeptics, I'm wondering what low-FODMAP antagonists have to say.
The problem with the FODMAP concept is that it's generalized to the whole food supply and population. So it would be as if because there are people with peanut allergies, we tell the

entire world to avoid peanuts. But, of course, peanuts are only toxic if you are sensitive to them.

FODMAPs are following the footsteps of gluten, in other words.

We guess that's what's coming. But, like gluten, you only need to eliminate FODMAPs if you are sensitive in the first place. So if you eat foods that contain FODMAPs, things like milk and honey and blackberries and garlic, and you feel fine, don't worry about it.

THE KETO DIET

Can we talk about the ketogenic diet? There's a lot of energy around that right now.

Ketone bodies are breakdown products of fat metabolism. The body makes them during starvation when blood glucose and the body's carbohydrate reserves (glycogen) have been consumed. The state can be induced with an extremely low-carbohydrate diet as well, since that prevents the body from replenishing glucose or glycogen and keeps it running on ketone bodies. Other than occurring as a natural consequence of starvation, the diet was first developed medically for the treatment of intractable seizures, particularly in children. It was also used to treat diabetes before insulin was available.

Treating diabetes and seizures—that's a big deal!

Yes, it's a big deal—but let's be careful. The ketogenic diet has been used as a treatment of last resort when seizures resist other efforts; that in itself doesn't recommend it to the general population. The ketogenic diet is high in fat and protein and low in carbs; eliminates fruit, grains, and beans, and many vegetables; and is high in meat and low in plant foods. That makes it the *opposite* of what we know constitutes a healthy diet, and one that's heavy in foods associated with adverse effects on long-term health as well as the environment.

Does the ketogenic diet still affect my brain if I'm using keto for weight loss or to help diabetes?

The ketogenic diet suppresses electrical activity in the brain even if your intention is to lose a couple of pounds. If you don't suffer from seizures, we doubt that's a good thing. Electrical activity in the brain is related to some fairly important stuff, like thinking.

How are Paleo and ketogenic diets different? It sounds like the foods they emphasize are the same.

Both emphasize whole foods and healthy fats while eliminating added sugar, grains, and legumes. But the keto diet focuses on shifting your macronutrient distribution toward fat in order to induce ketosis. It really de-emphasizes foods heavy in carbohydrates, and it allows for dairy.

Paleo doesn't have any restrictions on macronutrients, but de-emphasizes most dairy, legumes, and grains. Perhaps more important, the Paleo diet, whether it's practiced well or badly, is predicated on a robust idea: that the diet any species is adapted to eat is good and right for that species. That just flat-out makes sense.

The ketogenic diet, however, has been devised to mimic what starvation does to the body, and makes promises about rapid weight loss. But unlike a well-practiced Paleo diet, we have no evidence that a ketogenic diet is safe, let alone healthful, across a lifespan. That's very different. The two diets share an emphasis on meat, but otherwise should not be conflated.

Does eating bacon, steak, and butter really work for weight loss?

Yes, it works for short-term weight loss, as all highly restrictive diets do. The initial or "induction" phase of the Atkins diet is ketogenic, but it obviously does not work for the long term. If it did, the obesity problem would have gone away in the 1970s when that diet first became popular.

Do we know about long-term effects from the original Atkins diet?

Since tens of millions of people have had access to the Atkins diet since it was devised, we'd think that we would know by now if the ketogenic diet is the answer we have been waiting for; there would be tens of millions more of really thin people. But that's not the case: Rates of obesity have only gone up. So it's not clear that there are positive long-term effects among large groups of people.

In terms of health overall, there are no long-term studies of high-fat, low-carb diets, whether called Atkins, keto, or anything else. But since it is directly at odds with every aspect of diet we know *does* foster health across the lifespan, there is abundant reason to be dubious. Remember: We need to be skeptical about newfound "discoveries."

And let's be clear here: The burden of proof resides with those who want to claim, "This is good for you despite being at odds with everything we know to be good for you." The burden of proof does not reside with those of us saying, "We doubt it."

A diet that emphasizes red meat and creamy butter does appeal to me. Is there a more sustainable version of the keto diet?

The fat intake could conceivably come from eating a whole

lot of avocados, and there's a plant-based version of the At-kins diet called the Eco-Atkins diet, developed by research-ers at the University of Toronto. Such a diet could conceivably be used by those wanting to be ketogenic while still honoring concerns about health, animal ethics, and the environment. That doesn't allow you unlimited red meat and butter, though.

Might that version of the keto diet be beneficial?
Ketosis itself hasn't been proven to be injurious to health over the long term, but it hasn't been studied over the long term, either, and remember: Absence of evidence is not evidence of absence. Nor is this a common or "natural" diet, one you'd de-vise for yourself. Finally, it's a diet that denies you most of the really healthy foods you want to be eating more of, not less. In short: We are very dubious about its benefit, and aren't even convinced that it's safe. We hope it will be a short-lived trend, but time will tell.

But isn't weight loss a good thing?
The fact that biomarkers improve in the context of acute weight loss says nothing about whether the mechanism of that weight loss is good in the long term. *How you lose weight mat-ters.*

But *any* diet inducing weight loss in people who have typi-cal metabolic problems — high blood cholesterol, high blood lipids, high blood pressure, high blood glucose, insulin, in-flammatory markers, etc. — tends to improve those markers, whether the diet pattern itself is sustainable or healthy.

So you don't recommend the keto diet.
We have ample reason to be concerned about the ketogenic diet. It's hard to stick to and incredibly restrictive. It's at odds with everything we know that is associated with longevity and

lifelong vitality derived from diet. It's at odds with everything we know about planetary impact, environment, and sustainability, too. We vote no.

THE WHOLE30

How about the Whole30?
The name relates to the array of foods that the diet is emphasizing, and the fact that it's a 30-day regimen. Despite its momentary popularity, out of all the diets we are discussing, this one is the most readily dismissed. The annual *U.S. News & World Report*'s list of the best diets placed the Whole30 right at the bottom.

Why?
It invokes the tried-and-failed approach of every fad diet: Eliminate a lot, then add it back for livability. The Whole30 is somewhat Paleo-informed, and asks you to cut out a lot of foods. While it isn't necessarily marketed as a weight-loss diet, of course many people turn to it hoping they will lose weight fast.

Will I lose weight fast?
Probably. If you limit choice, you reduce the calories you consume. But then, gradually and incrementally, you add the foods back to your diet to try to achieve a diet that an actual human being could live with. And then you undoubtedly gain the weight back.

What foods does the Whole30 diet recommend cutting out?
All grains. All legumes. Many fruits. All dairy. Some of these, of course — whole grains, legumes — are among the most nu-

tritious, satiating foods, and the ones most closely associated with both finding health and losing weight. It's a ridiculous notion.

Dietary Patterns and Lifestyle

WHEN SHOULD I EAT?

Is breakfast really the most important meal of the day?
No. There is nothing holy about breakfast (unless you pray!).

Really? I'm a little bit in shock.
Take a breath. We realize it goes against all the folklore you've been fed since you were a kid getting ready for school.

But that's what it was—folklore? Breakfast doesn't matter?
Nope. There is nothing special about breakfast.

So I can just skip it?
I mean, you are technically going to "break your fast" every day at some point. But it doesn't have to be the moment you open your eyes in the morning, and it doesn't have to be with "breakfast food." Whatever you eat first, whenever that is: That's breakfast.

So I don't have to eat right when I wake up?
By and large, the literature that took on a life of its own was about kids going to school hungry. Not surprisingly, kids who wanted to have breakfast but had none were distracted in school. (To the kids who didn't care whether they ate early in the day, it made no difference.) From that grew the story that breakfast is the most important meal of the day.

Interesting. Well, is there an ideal time to eat my morning cereal?

Nope, nor does it have to be cereal. Some people are hungry first thing in the morning, some people are not. If you're not, and your energy level is good, you may find that forcing yourself to eat when you don't want to does more harm than good; eating when you're not hungry might increase the total amount you eat that day.

In short: When people eat is idiosyncratic, and that's fine. There are no rules regarding timing of eating.

So if I'm not hungry in the morning, it's okay—and even better—to simply not eat?

Eat only when you're hungry; don't eat when you're not. One of the great advantages of eating when you're hungry is it starts the day by essentially reasserting the importance of that comfortable, natural relationship with food. It's not about some arbitrary set of rules that you typically get with a diet.

Honestly, my work schedule determines whether or not I have time for breakfast more than anything else.

That's fair; routines differ. If your morning is extremely busy and trying to fit in a meal is hard to do, wait.

That's awesome that when I eat is totally up to me and my schedule. But what should I eat?

Myths abound about what's best to eat for breakfast. For example, some will tell you it's really important to start the day with protein — or a high-quality carbohydrate. But around the world, the food that's eaten at breakfast comprises just about everything there is: Vegetables, dairy, whole grains, fruit — they're all fair game. There doesn't appear to be any one best way to do this.

Are you sure? There isn't a specific type of food that's better to start off my day with? Oatmeal for fiber, eggs for protein?

The same high-quality foods that are good throughout the day are good at any given time of the day. You could even eat leftover dinner for breakfast. As long as you have the same sensible assembly of wholesome foods, the sequence doesn't matter at all. Eat your favorite breakfast. Needless to say, it should not routinely be pancakes, bacon, and maple syrup.

So when should I be eating my biggest meal?

There are all sorts of arguments about this, and there are some new studies that suggest that there are some benefits to the old recommendation that you should front-load your days with calories.

But that's back to must-eat breakfast? Breakfast like a king, lunch like a prince, dinner like a pauper?

Exactly. The argument is that it's important to have fuel in the tank when you are actually running the machine.

Dinner is a social gathering, though; it's my favorite meal.

That's a great point and not to be overlooked. We wouldn't be too slavish about this. Many cultures prioritize dinner over other meals, and there are certainly social advantages.

Look: As long as the net eating you do over the course of the day is appropriate in quality and quantity, it seems to all come out in the wash. There's something that matters more than *when* you get a majority of your calories, and it's obviously how much total eating you are doing and the quality of the food you are eating over the course of an average day. That's what matters most.

ON VARIETY

Does variety matter?

That, too, is a matter of taste. One of us eats the same breakfast pretty much every day: either a multigrain cold cereal or steel-cut oats with a mix of berries and whatever other fruit is in season, sometimes with added walnuts. The other never eats the same breakfast twice in a week, and might start the day with stewed vegetables, or a load of fruit, or oatmeal, or *pasta e fagiole*. Eat what you like.

Is this back to "Eat a variety of foods"?

Sorta. But this isn't just about breakfast, and the variety thing is interesting. The American Heart Association recently issued an advisory on the topic of food variety that was a bracing reality check, effectively saying that the notion that you should "Eat a wide variety of foods every day from all the different food categories" is a problem.

Why is it suddenly a problem to eat a variety of foods?

It's a great thing if it means eating brightly colored fruits and vegetables. But otherwise, there's plenty of room to misinterpret "wide variety." That's not very specific, and could be read to mean "eat everything." And "everything" isn't that good. In the 1970s, the typical supermarket in the United States had an inventory of about 15,000 products. In 2018, the typical supermarket in the United States had an inventory of about 50,000 products. How many more fruits and vegetables do you think we have today?

Not very many.

Exactly. So essentially there has been a mad proliferation of stuff in bags, boxes, bottles, jars, and cans. And that is *pseudo* variety, because overwhelmingly all those different products are made from the same few ingredients: soy, wheat, corn, rice, sugar, and some kind of oil. It's the same ingredients over and over and over repackaged into all these different products, and you'd be fine never eating any of them.

So, breakfast variety might be a muffin one day, the next a donut, and the following a Danish. All of this can be construed as "variety," but it's the same ingredients put together in slightly different ways, and they're all pretty much ultra-processed junk; the same is true at lunch and dinner. When a lot of the variety is not a variety of foods, but a variety of food products made from the same array of components, you are not getting the benefit of diverse nutrients. In fact you're eating the same junk over and over.

And this is just entirely different from variety when it comes to fruits and vegetables.
Entirely. We should say "eat a variety of natural foods," in the sense that "natural" means "close to nature." Fruits and vegetables are members of families that total to hundreds and even thousands of nutrient compounds. While all fruits and all vegetables have some portion of them, not one of them has them all. So in the case of plants, grains, nuts, and seeds, variety is important because that's how you are going to get all the nutrients you need.

What do these magical nutrient compounds do?
We don't know exactly. That's the beauty of it. They make you healthy. Good enough?

How does eating a variety of foods affect how full I feel?
Our appetite center is in the hypothalamus, and it's subject to "sensory specific satiety." It's why you always have room for

dessert: You're full for most flavor categories, but you haven't reached your threshold for sweetness. (See page 171 for more about your sweet tooth.) That's not necessarily a good thing, but as long as you keep the foods in the good-for-you category, it helps you enjoy eating.

So including different flavors and spices in our diet actually makes us feel fuller?
Spice tickles the appetite center. It's long been known to neuroscientists that we fill up in a sensory-specific or a flavor-specific manner. So if you keep eating the same thing, no matter how good it is, you lose interest in it. If you switch over to food with a tasty, different flavor, you rekindle your interest anew.

So you're telling me that our taste buds are designed to enjoy interesting flavors and delicious food? That's awesome.
It's awesome on one hand, but it also allows the food industry to manipulate us.

How does the food industry use flavors to manipulate us?
Basically, ultraprocessed foods squeeze a wide variety of flavors into individual food products. The idea is to get sugar, salt, and fat — foods we're hard-wired to eat, and foods that make us overeat, either because they're high-calorie (fat) or addictive (sugar) — into everything possible. If processors put as much salt into breakfast cereal as goes into salty snacks, and as much sugar into pasta sauce and salad dressing as typically goes into ice cream toppings, our appetite centers arrive at a state of super excitement. It literally stimulates overeating.

Okay, that does sound majorly manipulative.
Big food companies design food we can't stop eating, both

because of the variety of foods and also because of the variety of flavor stimulants in the individual foods. They're taking advantage of our sensory stimulation to make us eat more of their products.

SNACKING

What's the role of snacking in all this?
Snacking could be good or bad. When we were kids, we were told not to snack between meals because it would spoil our appetite; but "spoiling" your appetite in an age of overeating is not necessarily a bad thing. It's impossible to go wrong by eating a piece of fruit or a cucumber or carrot, no matter when you eat it.

You're saying it's what I snack on that matters.
Snacking is there when you need it. You're peckish. Maybe your energy level is starting to slip a little bit. And then, an hour and a half or two hours after you have a snack, when it's lunch or dinnertime, you're less hungry than you would've been. Well, that all sounds good, right? As long as it's a healthy snack, it is.

But is there any evidence that shows the benefit of snacking? Other than my own research that it puts me in a better mood when that three p.m. slump hits.
We do have evidence that nutritious snacks are good for you. Remember, foragers routinely graze — they snack all the time, eating what they can when they can. While we have established breakfast, lunch, and dinner as the modern norm, it's merely convention, not biology.

I don't want to stress out about choosing the wrong snack, or eating when I shouldn't.

You've got to be selective about your snacking, but it doesn't have to be that hard. As in other contexts, it helps to be prepared. You bring an umbrella when you think it might be rainy. In the same way, we live in a climate where unhealthy food is everywhere, so just as you would bring an umbrella to stay dry on a rainy day, bring along a healthy snack so you aren't tempted to go to the vending machine.

A good snack versus a bad snack . . . do I have to ask this question?
Apples, walnuts, bananas, carrots, hummus, bean dip, salad, etc., are all good. You know what we're talking about. Snack on high-quality foods all you want.

> Just as you would bring an umbrella to stay dry on a rainy day, bring along a healthy snack so you aren't tempted to go to the vending machine.

On the other hand, a snack from most glow-in-the-dark vending machines is almost certainly a food that's been engineered to put your appetite center into overdrive. You will wind up feeling crummy because you will get a spike in blood sugar; you'll be extra hungry later because this is the type of food that causes fluctuations in hormone levels that actually increase total eating. Especially if you grew up eating candy bars, you're going to crave them; but you know they're not what you need.

So it's a balance between following a healthy diet and also what works for my body and routine?
There are the fundamentals that pertain to us all: wholesome foods, healthy combinations, balance, reasonable variety, and all that. But exactly how you get there from here varies highly from individual to individual. Snacking is one of the great opportunities to personalize the format.

What should I eat when I'm not in total control of my routine, like when I'm traveling?

It's the same as snacking: Travel with wholesome foods you can carry with you. Think about creating a personalized snack pack: You know best what can work for you.

It's also a little easier than it was: As word gets out about simple, wholesome, minimally processed foods, they're available in more places. It used to be it was impossible to find a banana at an airport. That's not the case anymore. Hummus and guacamole are everywhere.

How do I make healthy choices at a restaurant?

With the same basic principles in mind: less processed food, and more emphasis on plants. You can order two soups; you can order a soup and a salad. And so on. But remember we're not talking a "salad" of cold cuts, cheese, and white bread croutons. Even if it's called a "salad," remember that what you call what you swallow doesn't matter; what you swallow does.

EATING LOCAL

What about trying to eat locally?

"Locavore" is a relatively new term but an ancient way of eating. Locavores eat — or try to eat — locally produced food, although of course the word "local" is itself pretty vague. Still, before food was shipped, everyone was a locavore. Now it's come to mean "what you eat matters to your health, and the quality and composition of what you eat is in turn determined by how the food is raised, what it's fed, and where it's from."

Why is it worth spending extra time and, usually, money to eat locally?

Reducing carbon footprint, supporting local economies,

eating seasonally (and fresh), knowing where your food comes from and how it was raised . . . all these are inarguably positive attributes, and all are characteristics of local food and pretty much *only* local food. No one but a fanatic could eat *only* local food, but concentrating on these attributes would mean you're eating better, more ethically, more sustainably — and more locally.

So you're telling me it's more complicated than just eating more fruits and vegetables. There's a whole other layer of what the produce is "eating" and how it's grown?
Yep, we are saying that. But before saying it again, and expounding, let's note this: Don't make perfect the enemy of good! It's good to eat more vegetables and fruits, even if they come from far away. It's just less good than if they are grown locally. We all need to do the best we can in the real world, and the best we can is rarely perfect. There's not a lot of "local" olive oil or coffee or bananas in the United States.

Do sunlight, rain, and soil quality affect how healthy an apple is?
Absolutely. Better soil means plants with better nutrient content. And that's not even talking about pesticides. For example: A famous study in the Linxian county in China showed that selenium supplements slashed rates of esophageal cancer. Why? The soil there is low in selenium, so the food grown there was, too. Selenium deficiency was widespread.

If you know your produce is being grown on a local farm where chemicals are not being used, you know that you are avoiding those chemicals. If you know the soil is being nurtured appropriately, then you know the nutrient composition of that soil is enriched. If your food has not been in storage and transit for days or weeks, it will have much more of its native nutrient content when you eat it.

So local sourcing means better nutrient composition of the foods?

In general, yes. (If you're sourcing industrially produced corn locally, then probably not.) It makes sense that agriculture and health are connected — there's no way around it. So the more you know about how and where plants are grown, the more confident you can be about their nutrient quality.

Does soil composition affect animal products, too?

Oh, yeah. This issue is even more important when you consider animal foods. The nutrient composition of meat is substantially connected with the diet and exercise pattern of the animal. Meat is leaner and less rich in saturated fat when animals get more exercise, for instance. The composition of meat also varies with what animals eat and the nutrients in those plants, which are in turn influenced by the soil.

Healthy, nutritious soil means healthy, nutritious meat?

When animals graze (on grass — the words have the same root) as opposed to grains, they keep the soil healthy *and* produce better meat. And pasture-raised animals have lower risks of industrial food-borne scourges, like *E. coli* 0157:H7, a strain that can cause severe infection and even kidney failure.

Where did this bad actor come from?

It came from the intestines of cattle, particularly those who were fed grain rather than grass! The industrial method of feeding cattle changes the pH of their GI tract, which creates different growing conditions. This alters the microbiome of the cattle and makes it more hospitable to problematic bacteria.

So a dangerous strain of *E. coli* evolved because cattle were fed outside their native diet?

Right. So if animal rights and environmentalist arguments

don't motivate you to eat local, remember that it also matters to your personal health.

Could you consider locavorism a diet?
It's a diet — and a good one at that — because when you look at the common core of good diets, there is always an emphasis on real, whole, high-nutrient foods. These foods are ideally produced free of pesticides, pathogens, and other contaminants (to the extent possible in this world) and in well-maintained soil. All these properties are best addressed with local, nonindustrial food production.

Having said that, it is of course possible to produce bad or nutritionally deficient food locally. But a main advantage of sourcing locally is that you can see or learn about the production and easily make that judgment.

It's also pretty safe to say that a locavore diet does not include nationally distributed food products made in a factory somewhere, which pretty much means it excludes junk food.

Does that mean I can eat whatever I want? Grass-fed burgers and locally produced cheese all day!
Not exactly. It's still important to think about a plant-predominant diet — along with balance and variety. The closest approximation to locavore thinking translated into actual dietary patterns would be a flexitarian diet (more about that on page 47).

Foods and Ingredients

What do you mean by the terms "whole food" and "real food"?
It's pretty simple: food that's as close to "natural" — that

is, as nature produces it — as possible. We know that ultra-processed (or hyperprocessed) foods are bad for us: That's the majority of food invented in the twentieth century — junk food. Whole food is generally food that hasn't been tinkered with much, that doesn't need a label, and is its own ingredient. (Although too much meat and dairy can be harmful, in their simplest forms, they are whole foods, too.)

Okay: How do I eat enough to make sure I get the nutrients I need?

Since the food supply is booby-trapped to make us all over-eat, like most Americans you're likely eating more than you need. If we told you that you ought to add twenty-seven kinds of foods to make sure you got enough of some micronutrient or other . . . you'd just be eating more. That's not the way to go about things. We need to be eating more of the right foods *instead* of many of the foods we're eating now.

Focusing on the balance of food makes it easy. We know which wholesome foods are associated with health. Focus on those, and you wind up in the sweet spot regarding nutrients.

How do I make room for more healthy food in my diet?

You don't have to eat less overall, you've just got to eat less of things like cheeseburgers, chips, and ice cream. This is a fundamental truth, and we'd be shocked if you didn't know it already. If you're eating the standard American diet (SAD), you just have to give up some foods to make room for others. If we give up the foods that we eat too much of — generally speaking, ultraprocessed food, junk food, and industrially produced meat — and eat more vegetables and fruits (we know — blah blah blah) — we suddenly have a near optimal diet.

When you are thinking about reducing your intake of unhealthy food, also think about what you are making room for.

It seems obvious, but you are always going to get 100 percent of your calories from all the food you eat. To use one example, one reason we tell people to eat less beef for the sake of personal health (let alone the health of the planet) is because the more beef you are eating, the less you are eating other, more beneficial foods that could take that place in your diet, like beans, lentils, or fish. If you replace beef with cheese, you may be "becoming a vegetarian" but you're not doing yourself any good. If you replace beef with lentils, you've got a real winning situation.

The typical American diet is not beef deficient. But the typical American diet *is* deficient in fruits: vegetables, beans, lentils, nuts and seeds, and whole grains, plant foods that are most reliably associated with net health benefit. The more you eat animal products and junk, the less room you leave in your diet to get your calories from these food sources. So: When you are thinking about reducing your intake of unhealthy food, also think about what you are making room for.

When you substitute one food for another, there's always the question of "Instead of what?" If eggs replace donuts, they are good; if they replace berries and steel-cut oats, they are bad. It's all relative.

But I like ultraprocessed food. And a burger is much more satisfying than a green salad.
We get it, but even though ultraprocessed food is designed to increase consumption, even though eating it — per a recent, solid randomized controlled trial — leads directly to overeating, replacing it with more wholesome options is not as hard as it might sound, for two reasons: First, wholesome foods are generally quite satiating (okay, maybe not a green salad, but stand by), meaning you will want less junk food, and second, as you acclimate to more wholesome food choices, your taste buds go through rehab. They quickly learn to prefer

those more wholesome foods, and you lose your taste for junk. We can personally attest to this, as can others: It really happens.

What's the easiest way to make sure I'm eating a balanced diet?
At the end of the day, the best thing to do is to eat those foods that are recommended for health: The ones we mention above make up a complete list. Those foods provide ideal distributions of nutrients. Eat a variety of wholesome foods, and the nutrients will take care of themselves.

FRUITS AND VEGETABLES

Why are fruits and vegetables so important?
Fruits and vegetables are the foods that are closest to nature, easily found in their most unaltered, natural form. As we say elsewhere, all of the world's best diets are plant-based, so you can't really go wrong with fruits and vegetables. Eat them. Eat them all the time. Eat them to the exclusion of animal products and junk.

Is drinking green juice with four servings of vegetables and fruit the easiest way to get this done?
There's actually a huge difference between drinking a juice and eating unprocessed, whole produce.

Something that can be consumed in a gulp rather than by multiple slow chews changes the equation. Most juicing, especially that done commercially, strips away the fiber and many of the beneficial nutrients in produce. Juices tend to have high glycemic load, which means a) they contain sugar that gets quickly into the bloodstream, and b) they are a concentrated source of that sugar. (See page 172 for more about glycemic load.) The sugar content is the same in the unprocessed food, but it enters the bloodstream much more slowly, with the benefit of lots of fiber, and that does matter. Keep in mind, too, that you can drink the juice of four apples way more easily than you can eat four apples, and — for all the reasons above — that's not a good thing.

So juice cleanses really aren't all that they are made out to be?

Unless you're doing them instead of eating cheeseburgers, fries, and sodas, they're nonsense. (If you are, they're an upgrade, but you need to rethink your diet.) Again, remember: "Instead of what?" If your diet is good to begin with, you don't need a juice cleanse. If it isn't, and you want it to be, a "juice cleanse" could be a good jump-start; but don't expect it to be anything more than a "cleanse" of bad habits and the start of something new. The notion that it will wash poisons out of your body is silly.

What about smoothies?

Again, if you don't eat fruits and vegetables routinely, but you're willing to blend them into a smoothie and drink them, that's almost certainly going to be a benefit to your overall diet. And if you blend the whole fruits and vegetables yourself, that's definitely different from buying an already-extracted juice, which is probably robbed of fiber. Using a blender for produce keeps the pulp, or fiber, in the mix, which does blunt the glycemic response, as noted above.

Still, better to eat the foods rather than drink them.

The glycemic load of a smoothie — the speed at which its sugar enters your bloodstream, and its effect, first on blood sugar and then on blood insulin levels — is higher than if you slowly ate those same fruits and vegetables. Also, chewing a solid food takes longer, further slowing the entry of sugar into your bloodstream. And finally, solid foods fill us up more than liquids. So: Whole foods are better, and eating fruits and vegetables is simply the best thing you can do. If those aren't an option, blending them is a fine fallback position. A smoothie made of good stuff you wouldn't otherwise eat at all is a terrific example of not making perfect the enemy of good. Just don't fall for the "Green Goddess Cleanse" at the store.

Can we talk about the difference between organic and not? Like, does any old apple a day keep the doctor away? Or does it have to be organic?

Let's start with the question about organic that's most relevant here: Does it confer dietary human health benefit? The answer is, honestly, "That's impossible to say right now." (That organic practices are beneficial to the soil, to farmworkers, and to animals is unquestionable.)

Why is it impossible to say?

First of all, almost nobody eats purely organic versus purely nonorganic, so it's impossible to divide people into two groups for comparison. And on the flip side, it's not like anyone goes out of their way to refuse to eat organic. So there's an overlap, and when two groups that are supposed to be different are a bit alike, their outcomes look more like one another.

Isn't organic food purer than all the other stuff that's buffed up with chemicals?

Purer, yes, but not "pure." You can raise food organically, but you are still obligated to use the water that exists on this planet, and that's been contaminated to one degree or another. So even organic food isn't perfectly spared the same exposure as food grown conventionally. Perhaps more important, in the words of our esteemed friend and colleague Marion Nestle, "Organic junk food is still junk food."

Buying organic gives me peace of mind, though, knowing my food contains no added chemicals.

We can't really routinely measure herbicide and pesticide content in food, but yes, it's safe to assume that certified organic food contains fewer harmful chemicals. That's a decided upside.

If I don't buy organic, does washing my fruit and vegetables help reduce the presence of herbicides and pesticides?

Yes. Rinsing conventionally grown produce reduces the pesticide residue, and so arguably narrows the difference between organic and not organic.

Although the assumption that organic is better than conventional has always made sense because fewer synthetic chemicals are used, scientifically confirming that sense has been nearly impossible, for the reason we mentioned above. But again, organic agriculture is safer for soil, farmworkers, and animals at the very least — and probably eaters, too.

And this *is* becoming more clear: There is a burgeoning array of large observational studies that look at people who reported willfully eating organic routinely versus people who didn't. As of this writing, the most recent study is out of France, where researchers found a significant difference in cancer incidents between those who eat organic routinely versus those who don't. (Those people who ate organic most often had the least cancer, as you'd expect.)

Well, doesn't that mean that organic is better after all—if it prevents you from getting cancer?

The study really isn't conclusive, though.

Why not?

Maybe the result is because people who intentionally eat organic routinely differ systematically from people who don't.

Maybe what makes the two groups different is the total level of care they take with their health. It could be that all those people eating organic have access to better medical care, more cash to spend, higher quality of life, that kind of thing. Generally organic food costs more, which means more well-to-do people, those with better healthcare and usually better living situations, are more likely to be the ones eating better. So we still don't know for sure. But at least for the first time we have a strong association between routine consumption of organic food and an important health outcome.

So the focus on organic leaves out a lot of the population?
For sure. One of the concerns about overemphasizing the value of organic foods is that you may be talking to a very narrow sliver of the population. And that's a problem, because if people hear this message that fruits and vegetables are "contaminated" with pesticides, they may think that if they can't buy organic, they should avoid these foods entirely. So sometimes we actually talk people out of eating produce altogether because we've convinced them that they are delivery vehicles for pesticides. Not good.

Kind of in the same way people avoid fish because of the toxins?
Kinda, but this discussion about produce holds a lot more weight; it's more important. The net effect of eating produce overwhelms the potential liabilities of whether it was grown conventionally or organically. A conventional apple is still a better food than an organic snack bar.

If I'm at the grocery store and am looking at organic apples and nonorganic apples side by side, and the organic apples are too expensive and I can't afford them, what should I do?

A nonorganic apple is better than no apple, and better than most other choices. Rinse or wash it well to minimize the chemical hitchhikers. It's almost safe to say, "Never pass up an apple."

And if there's an organic option, and I can afford it?
If you can afford to buy organic, that's the better choice. But it's important to remember that it's not the only choice.

Should I buy all organic produce if I can? Or do some fruits and veggies matter more than others?
You can refer to the Environmental Working Group's annual report on the fruits and vegetables that contain the most pesticides, and start there, but if an organic option exists and you can afford it, go that route. If not, eat fruits and vegetables anyway; just wash them first.

WHOLE GRAINS

Are there health benefits of eating whole grains like oats instead of packaged breakfast cereal, or brown rice instead of white rice? Wheat bread versus white bread?
Products made from whole grains — intact whole grains, in minimally processed form, which excludes commercial products like "wheat bread" — are reliably better for us. We are almost all deficient in fiber, and that's in large part because we eat ultraprocessed grains, from which the fiber has been removed.

Be careful of this difference: Made WITH whole grains is not the same thing as made FROM whole grains. The former means there might be some whole grains in there; the latter means it's a whole-grain product. "Wheat bread" is usually made with mostly white flour, or at best half and half, and in most products even the whole wheat contained therein is questionable.

Remember — traditionally, bread has been a whole-grain product. Wonder Bread is a product of the early twentieth century.

The short version here is, eat whole or minimally processed grains, and don't be fooled by marketing.

That's tricky. How do I know which grain products to buy?
Use these rules: The simpler the products made from grains, the better they are. Whole grains are best: brown rice, wheat berries, cracked oats, farro, quinoa . . . literally whole or nearly whole grains. Whole cereal, like oatmeal, is great.

Then beware of misleading labels: Look for "whole" in the ingredient list; if it doesn't declare it, it's probably not so. "Multigrain" does not mean "whole grain"; "wheat" does not mean "whole wheat." Processed grains that are not listed as "whole" are mostly worth avoiding.

Two more things: If the product is labeled, look for 2 grams or more of fiber per 100 calories; that's a good sign. And check the other ingredients: The best bread contains ground whole grains, water, leavening, and salt. The best cereal contains ground (or not) whole grains, period.

Does cooking whole grains do anything to their nutrient composition?
When you cook them at home, probably not. Sometimes cooking reduces nutrients, sometimes it concentrates them; it's not worth worrying about, and certainly the fiber is unaffected. However, when high-carbohydrate foods are cooked at very high temperatures in industrial food production, they may create a compound called acrylamide, which has been identified as a potential carcinogen.

Wait, whole grains can give me cancer?
We didn't say that! Don't freak out. First, this is not a liability
particular to whole grains — it is true of foods that happen to
contain both carbohydrates and a particular amino acid, as-
paragine. Such foods include most grains, potatoes, and coffee beans, among others. Acrylamide can form in such foods
with high-temperature cooking. Trace amounts of acrylamide
in a food that's otherwise good for you are not worth worrying about. Coffee, for instance, is a source of acrylamide — but
routine coffee intake is associated with less cancer, not more.
If the food is junk in the first place, like many chips and salty
snacks, then acrylamide is just one more reason not to eat it.
Ultraprocessed food products are worth avoiding, for a variety
of reasons. The overall composition of a food, and the overall health effects of eating it, almost always matter much more
than the theoretical risk or benefit of some isolated element
in it.

**A lot of my friends don't eat gluten, but I admit I don't
actually know what it is.**
You're not alone — many people are afraid of gluten without
knowing what it is.

Well, what is it?
It's a complex protein.

It is? Is it new?
No. It's been around as long as wheat, and wheat is one of the
oldest of the foods grown by humans, beginning in the earliest
days of agriculture more than ten thousand years ago.

Did this ancient wheat contain gluten?
Absolutely. Humans have been consuming gluten routinely
for millennia.

Then why does it seem like gluten intolerance is a new thing?

Because everyone is talking about it. When we freak out about something, we tend to see it everywhere. So you know if you occasionally have a headache, you are much more likely to blame it on gluten after everyone is talking about gluten. If you have an upset stomach, you're much more likely to blame it on gluten because you've heard others do so. It's a convenient scapegoat.

So it's all in our heads?

No, it's not that easy, and we're not saying that. There is also the fact that there really does seem to be a rise in gluten sensitivity and the more extreme form of it, known as celiac disease (or gluten enteropathy), when the body makes antibodies to gluten.

So gluten is bad for you. Has it always been a terrible thing and we are just now realizing it?

Nope, that's not the answer. Most people are gluten tolerant and just fine. People in many cultures have been thriving on whole wheat and whole barley — which also has gluten — for thousands of years.

Has the rate of celiac disease actually gone up?

Maybe. About 1 percent of the population has celiac, and that's been pretty stable, but it might be rising. The rate of people who are gluten intolerant or who claim to be helped by avoiding gluten is as high as 10 percent. That means, of course, that about 90 percent of the population can consume gluten with no adverse reaction. It also means we're way more aware not only of celiac disease but of gluten intolerance, which is milder than celiac but also in many cases legitimate. Why this has happened, we're not sure — although changes to the microbiome related to changes in the food supply are a likely factor.

Is gluten different now than it was ten thousand years ago?
Gluten is the same as it ever was. But varieties of wheat have changed, the way they're grown and processed has changed, and the company that gluten keeps in food has changed, too. We've hybridized wheat thousands of times, loaded it into ultraprocessed foods, and added many chemicals to the mix.

Sometimes a compound that does not trigger intolerance on its own can do so in the company of another compound. That may be part of the reason gluten sensitivity appears to be on the rise: more hybridized and otherwise altered products, more food processing and chemicals, and more disruptions of the microbiome, which can alter how the GI tract reacts to almost anything. An unhappy microbiome could very well mean an inability to digest gluten — and other things — properly. (See page 189 for more about your microbiome.)

So those things are the source of the gluten-sensitivity problem?
We don't know what "the source" is. All the factors just noted are potential contributors, and the more researchers look, the more potential factors they find. In epidemiology, this is called surveillance bias: Problems arise as you look for them.

A change in the wheat crop seems like an easy explanation for why more people are sensitive to gluten.
Some claim that modern wheat is more concentrated in gluten than older varieties are. But some ancient grains contain just as much gluten as some of the newer grains. And besides, when it comes to antibodies, dose is substantially irrelevant: If you are allergic to something, even a tiny exposure will trigger that response. Concentration of gluten is unlikely to be a ma-

jor factor in celiac disease. But as we've said, it may be something else in modern varieties of wheat that triggers a gluten response. We do know — since you're bound to ask — that it *isn't* genetic engineering, because there's no commercially grown genetically modified wheat. All we really *know* is that there's increased sensitivity. The underlying reason(s) is/are not clear.

What's your best guess?

The most likely reason for a rising rate of gluten intolerance is not so much the gluten itself, but that we have damaged the native integrity of our gastrointestinal tract and microbiome, and disrupted our overall diet. Things that were not injurious now have the tendency to be — and things that may be injurious, from antibiotics to chemicals, are now routinely part of our food. And that's not just true for gluten, but for many other compounds as well.

So my headaches and stomachaches after I eat gluten . . . might not be from gluten at all?

They might not be, but sensitivity among one in ten of us is far from rare, so don't dismiss the possibility, either. If in doubt, get checked out. Because some highly nutritious foods contain gluten, it would be a shame to banish them from your diet if gluten is not the problem. But if it is, you may feel much better by giving it up, and there are alternative, wholesome, gluten-free foods, including other whole grains.

If I'm in the 90 percent of people who can happily digest gluten, will it benefit my overall health to eat gluten-free?

No. Anybody in the 90 percent who's giving up gluten because they heard it is a bad thing is wrong. There is actually a significant potential to degrade the overall quality of your diet by giving up gluten, because gluten is present in whole wheat, and barley, and other whole grains, too, and avoiding whole grains makes it difficult to get enough fiber. The typical American is fiber deficient (researcher Denis Burkitt once famously said: "By global and historical standards, the entire U.S. population is constipated") and that's only gotten worse over time.

There's another problem, too: There's a LOT of gluten-free junk food that exploits the current preoccupation. If you give up gluten for no good reason — just because you think avoiding gluten is "good" — you become vulnerable to the sales pitches for that whole new way of eating badly!

Some people can't tolerate gluten. But gluten isn't a bad protein, just as peanuts aren't a bad food — some people are just allergic to them. Avoid gluten only if there is a good reason to; but otherwise, eat whole grains, including those with gluten.

BEANS

Didn't I hear recently that beans are poisonous?
Beans are the most important source of protein in the world and among the best things you can put in your body. This latest conspiracy theory about diet — that most of the most nutritious foods contain members of a large protein family called lectins, and that lectins might be toxic — comes from a recent bestseller called *The Plant Paradox*, which is based on some cell culture studies and some animal research.

"Might be toxic" sounds pretty bad.
That's clickbait. Some studies suggest that lectins in raw beans may be toxic, but the reality is that nobody eats dried beans raw; you cook them. And then they're not toxic, but beneficial: Beans have anticancer effects and may help reduce the risk of chronic disease.

What does sense tell you? Beans are the world's most important protein, a food humans have eaten for at least ten thousand years, a staple in the world's healthiest diets, and a good source of almost infinite nutrients. Is that reason enough to eat them, or is some lone guy saying that they're poisonous going to scare you off?

So the consensus on beans is . . . ?
People who eat more beans, across the board, tend to have better health outcomes than those who don't. Beans are stunningly rich in a wide array of nutrients, stunningly rich in fiber. And that's backed up by pretty much every kind of evidence we could hope for.

How does adding beans make for a healthier diet?
Lots of ways. It adds fiber we need. Beans are rich in vitamins and minerals, and many are rich in antioxidants as well. And since beans can easily replace meat, eating them will likely mean eating less saturated fat.

In general, when eating more of A (say, beans) means eating less of B (say, beef), and a health benefit results, you might ask whether it's because of more A, or because of less B? And the answer is: Yes — likely both.

Are we supposed to just forget about the lectins?
Yes. Beans contain lectins. So do whole grains. And fruits and vegetables. Many of the healthiest foods you can eat contain "might be toxic" lectins. You have to reject the ridiculous argument that beans contain this potentially toxic compound and therefore everything we knew about beans must be wrong; that is a complete misapplication of science. (See page 197 for more on the limits of science when it comes to food and health.)

> Any diet that relies on beans instead of meat is going to be healthier than any diet that relies on meat instead of beans.

What about soy? It seems a bit more controversial than most other beans.

It should be an easy conversation. Soybeans are a bean, and an especially high-protein one at that. And beans are good. So — we're done.

But soy has turned into so many things, and there are so many forms of the soybean now that it's become a complicated discussion.

Well, a lot of soy is GMO, right? Does that make it more of a danger?

People who are concerned about GMOs (genetically modified organisms) often argue you can't eat soy because GMOs are bad. But genetic engineering is a method that could produce stuff that's good for us; it's still controversial whether or not it has yet. That's a long discussion, but the problem with GMO foods is not the technology but rather what's been done with it.

So whether soy is GMO or not isn't really the issue, at least nutritionally; what's done with that soy is. Most soy grown in the United States goes either to feeding animals or to producing junk food. (The same is true of corn.) A GMO soybean eaten in bean form will likely have the same positive impact on your health as a non-GMO soybean.

However: The prevailing problem with GMO crops at the moment is not the composition of the foods they produce but the treatment of the crops. A lot of the GMO activity is all about making crops tolerant of the herbicide Roundup (glyphosate), so that fields can be doused with it. Glyphosate is toxic to people and animals. (So, of course, are other pesticides.)

Is there other environmental damage from soy?

Soy crops have a big environmental impact because there are

vast seas of monoculture soy plants. (Same with corn. And wheat. And cotton. And rapeseed. And so on.) Monocultures are a problem largely because of their damage to the environment and public health — fertilizer runoff, pesticides and antibiotics entering the food and water supply, land degradation, and, of course, because the products are converted to ethanol, feed for confined animals, and junk food.

Okay, so what are the health effects of soy?

Are we talking about pure soy? Soybeans, edamame? Or are we talking about soy *stuff*, like soy hot dogs or "chicken"?

Well, isn't a soy hot dog better than a meat hot dog because it's vegan?

That's just ultraprocessed junk that happens to be made with soy. It might be "better," but that doesn't make it good. Vegan food can be junk food.

All right. Does adding soybeans, in the form of soybeans, to my diet make me healthier?

Yes: Research shows that people who routinely eat soy show a consistent health benefit versus people who don't. That's not surprising, because soy is a legume. Even traditionally processed soy products — tofu, tempeh, miso — seem to be parts of a good diet. And here you have the additional notion that tofu and tempeh are being used specifically as meat substitutes. As we've asked before: What are you eating it instead? Diets that are rich in soy tend to be lower in meat. That's a key point.

Where's the benefit coming from? The removal of meat or the addition of soy?

Probably both. It's nearly impossible to unbundle these effects, since they come together.

Could the same reasoning be used for other plant-based sources of protein, like chickpeas, pinto beans, and lentils?

Absolutely. Any diet that relies on beans instead of meat is going to be healthier than any diet that relies on meat instead of beans. When populations eating traditional diets start adding larger amounts of meat, their health declines. (As it happens, they're also tending to eat more sugar and junk food in general. The standard American diet is a bad package.)

Okay. But is there anything exceptional about soy in particular?

There is: Soybeans, unlike most beans, are highly concentrated in phytoestrogens, plant compounds that mimic some of the effects of estrogen, a hormone that contributes to healthier bones and healthier hearts.

But isn't there worry about phytoestrogens? What's that about?

There is concern about cancer promotion from phytoestrogens, based on concentrated soy nutrients in animal experiments. (This is also why there's debate about hormone replacement at menopause.) But these experiments are not the same thing as soy in the diet, which, as we've said, is beneficial.

So again, let's say you are eating more soy instead of bacon cheeseburgers, which weren't so terrific for your cancer risk, either. The net effect of that replacement appears to be beneficial. Remember, we're talking soybeans and minimally processed soy products, like traditionally made tofu, soy milk, and tempeh. That's completely different from soy sausage or ice cream, which may include twenty other ingredients; in those items, soy is just along for the ride.

DAIRY

As long as you bring up soy milk, let's talk dairy: good or bad?

Both, and in between. There are passionate arguments that dairy is good for health and that dairy is horrible for health. Interestingly, the most ardent answers in opposing directions are based on some of the same considerations. And some different ones. We'll get more specific, but first let's remind you of the question "Instead of what?"

What's that dairy replacing in your diet? And what might it replace? Eating dairy can improve a bad diet or degrade an optimal one. (In making these decisions, there are, of course, also the questions of how the poor cows are treated, and the environmental impact of keeping tens of thousands of cows on one farm.)

There are two main questions to consider: First, whether humans are adapted to consume dairy; whether it is a "normal" part of a human diet. And second: "Is all dairy the same?"

What do proponents of a no-dairy diet argue?

People who believe humans are not adapted to consume dairy look back to the Stone Age and say we're adapted to give up dairy the second we stop drinking our mother's milk. This is a robust argument, and it's not just about *Homo sapiens*, but about all mammals.

Is it true that after breastfeeding we can no longer digest dairy well?

One of the signature characteristics about mammals is that we drink our mother's milk. Maternal milk always comes with a sugar compound known as lactose; it's a complex sugar, a disaccharide. And in order for us to get our mother's milk into our bloodstream, we have to uncouple those disaccharide sugar compounds. To do that, we need an enzyme called lactase.

If we're born with an enzyme to digest milk, how come so many of us wind up being lactose intolerant?
We're all born with a gene that tells our bodies to make lactase, and we do. But then, in most humans, a funny thing happens on the way to toddlerhood: The gene that makes lactase turns off, and mammals lose the ability to digest lactose.

Why does the gene that makes lactase turn off?
It's evidently part of the preprogrammed instructions of that gene. It's only meant to go to work when this infant mammal is born, and when this infant mammal stops digesting its mother's milk, the enzyme turns off, with the assumption that it will never be needed again.

So that's why I can't eat dairy?
Some people can and some can't (those humans who cannot are in the majority), but evolutionarily, we were not originally adapted to consume dairy past infancy, and no one needs it. Throughout almost all of animal history, that enzyme would never be needed again because you wouldn't just encounter milk in nature. There's no dairy you just find in a puddle somewhere.

Do any other mammals consume dairy beyond infancy?
Some mammals *can* consume dairy beyond infancy, but only those that are "domesticated" — us, and our cats — and maybe some other animals in zoos.

It seems like this would be pretty hard to argue in the other direction if we look to evolutionary biology.

It does seem that way, but proponents of dairy use some of the same evidence.

They say that evolutionary biology and adaptation didn't stop fifty thousand years ago or fifteen thousand years ago, and that evolutionary forces are ongoing. If you are going to use adaptation to help determine what's good for people to eat, then what's good for people is going to change over time.

So we're evolving to drink milk?
Since the dawn of agriculture roughly ten thousand years ago in Mesopotamia, human beings have spread all over the world and encountered all sorts of different dietary exposures and stresses. When people spread north and colonized parts of the world with colder, more hostile climates (as the Vikings did, for example), they had to get creative with what they could eat.

The Vikings were subject to periods of significant shortfall of nutrition from plants during the long, dark winters of Europe's north. People obviously ate some meat; but they knew they would starve if they ate all the animals they could find. So they began to extract nutrition wherever they could find it.

They began to consume the milk of other animals.
Exactly.

So we first turned to milk not because we were curious or thought it was delicious, but because we were desperate?
Super desperate. A whole lot of human dietary diversity started with desperation. Consider being the first person to look at a lobster and say, "I wonder what THAT tastes like." Similarly: Can you imagine the initial person who consumed the milk of another species? That's desperation. (It's not strictly relevant, because it never spread the way dairy consumption did, but Mongols drank blood from their horses when necessary. Desperation.)

People were probably not thinking, "Wow, this milk is going to be delicious." Rather, they were really hungry, and figured, "If that milk can keep a baby goat alive, maybe it will work for me, too."

But these Vikings who were some of the first humans to consume dairy—they didn't have the lactase enzyme to digest it, right?
That was a problem, of course. Adult mammals don't make lactase, and all dairy has lactose. So their ability to digest, metabolize, and benefit from dairy was impaired. They probably experienced some digestive issues when they drank that milk; but digestive issues, you'll agree, are better than starvation.

So what happened in the long run?
Natural selection happened. If genes are giving instructions and genes are prone to mutations, then you end up with alternative versions of genes with alternative instructions. Somewhere along the line, and nobody can say exactly who or where, some Scandinavian — for the sake of this story, let's call her Gunhilde — had a mutated version of the gene with instructions to keep the enzyme lactase going past infancy.

When her tribe, let's say in Greenland, faced nutritional deprivation and everyone was trying to get by on whatever was available, Gunhilde was able to digest the milk that others couldn't digest nearly as well. Therefore, she derived greater nutrition from that milk and wound up healthier, and more likely to survive. She survived the famine that killed other members of the tribe and lived to mate another day. Her children, at least some of them, probably made lactase into adulthood, too — so they, too, wound up with a survival advantage. You can see where this is going.

Is that how natural selection works?

In part, at least. To put it more concretely: Natural selection works so that the genes that favor survival get passed along. People who survive are far better at making babies than people who don't survive. (Put another way, people who don't live long enough to make babies make for very poor ancestors.) So, in Northern European populations, and especially Scandinavian populations, the gene instructions changed and the prevailing pattern became, "Keep making the lactase enzyme." This, too, is evolutionary biology. This is adaptation.

But doesn't that sort of change take thousands and thousands of years?

Not always. Depending on the magnitude of the stress, evolutionary biology can play out very quickly. And in any case, this particular switch in this particular gene is either "on" or "off." When the adaptations are simple, and happen readily with a change in the "setting" of a gene (known as the epigenetic controls), it can happen very fast.

Specifics are impossible to know, and we are not anthropologists, but it's pretty safe to assume that for the Vikings, the likelihood of surviving a winter increased by a significant factor for those who could digest dairy efficiently.

Is whether I'm lactose tolerant dependent on if I'm related to those early Vikings?

Today, about 95 percent of people of principally Northern European descent — not just from the Vikings — are biologically adapted to digest milk throughout life.

And if I'm not of Northern European descent?

In contrast, if you look at populations that were never faced with those same survival pressures — most people whose ancestors are from Asia, Africa, and the pre-European-contact Americas — they are about 95 percent lactose intolerant. Other groups vary from one extreme to the other.

How do I know if I'm lactose intolerant?
Lactose intolerance is not rare and it's usually pretty obvious to people who have it. Most North Americans — far from all, but the majority — do a pretty fine job of digesting milk throughout life. And some people who are lactose intolerant can't drink milk, but can eat cheese, much of which has far less lactose than milk.

If I can tolerate lactose, how should I decide whether or not to eat dairy?
The net health effect of dairy consumption is one of the more legitimate controversies of modern nutrition. First of all, dairy is a whole category. There's milk, butter, cream (the concentrated fat portion of milk); there's dairy in its native state, where it tends to be a very concentrated fat source; there's the same dairy with the fat removed (so skim, fat-free); and then there is fermented dairy like cheese and yogurt, ideally produced with a variety of active cultures and bacteria, which are beneficial.

There's also a lot of moralizing around dairy, and some of it is justified. Vegans and animal rights advocates argue convincingly that dairy-eating is at least ethically challenging, if not wrong. And dairy production ranks high in its negative environmental footprint, with unsustainable associated water consumption and prodigious output of greenhouse gases.

And the health arguments are complicated. It's not something we want to be dogmatic about, and it's something we'd encourage you to remain flexible about.

Are there proven health benefits of including dairy in your diet?

There's debate about the effect dairy has on health, but it's pretty clear that there's no health benefit to including dairy if your diet is otherwise optimal; vegans do just fine without it, and dairy has some nutritional downsides as well. On the other hand, if you're drinking milk instead of soda, that's a trade-up. If you're eating minimally sweetened yogurt instead of ice cream . . .

But who really has an optimal diet?

Exactly. The whole thing gets more complicated when we use the backdrop of the generally horrendous typical American diet.

Are there no out-and-out benefits to drinking milk?

A number of studies have found that kids who routinely drink milk have fewer problems with weight. People in the dairy industry say this proves that milk is important for kids' health.

So is milk good for kids, the way it's always been said to be?

Instead of soda, or sugary fruit juices, milk is better; it's a nutrient-rich beverage, and pretty satiating (filling), whereas soda is nutrient-destitute, extremely sweet, and much more appetite-stimulating than satiating.

But children of vegans have even better weight trajectories than children who consume dairy, so you don't need milk to have a normal weight.

Is the fat in dairy good for me in the same way as the fat in avocados or walnuts or olive oil?

It's not. There's no evidence that dairy fat on its own is healthy, especially against the backdrop of an already well-put-together diet. But — to get slightly into the weeds — the fats vary among types of dairy, notably fermented dairy. There is ongoing research suggesting that we may need to think of the fat in cheeses and yogurts differently from the fat in milk. And we may even need to differentiate among cheeses. What's safe to say is that animal products, including dairy, should not be your *main* sources of fat.

What's the argument against including full-fat dairy in my diet?

It's a concentrated source of saturated fat, and — as we discuss later in this book — saturated fat has problems, and most of us get too much of it already. Again, by and large, the evidence would point against dairy as a healthy staple, but if we are placing it within the context of a really poor diet, it has advantages. Context really matters here.

Does fat-free, low-fat, or full-fat make a difference?

A lot of low- or nonfat dairy is really highly processed food. Skim milk is made by a simple physical process; you literally skim the cream from the top of the milk. But "fat-free half-and-half" is an abomination, and could not possibly exist in nature or even in a simply processed form; it's frankenfood. That's not a trade-up from anything.

What's the argument in favor of full-fat?

The satiating effect of full-fat dairy can be an important consideration. If you're drinking a glass of full-fat milk, it's probably more filling than fat-free. So if that glass of milk pushes some junk out of your diet, that's a good thing.

But you don't need the milk specifically; you just need to avoid the junk it might replace. If your diet is already rich in wholesome foods and the most healthful fats — from nuts, olives, avocados, seeds, fish — then there's no convincing argument that milk has additional benefit, and the fat in it could even be harmful.

Are you saying that the only health benefit of whole milk that is makes you feel full?

If your diet is made up of the most nutritious foods to begin with, you don't need to rely on whole milk. Just because you can tolerate lactose doesn't mean you *should* drink milk. It means you can.

Some say a lot of the nutrients in milk are present in the fat.

All the milk that's sold commercially in the US is fortified with vitamins A and D, and those are both fat-soluble vitamins, so there is probably a slightly better delivery of vitamin A and D when there's more milk fat. On the other hand, calcium is water-soluble — and fat-free milk, matched ounce for ounce, actually delivers more of that. So, too, for other minerals.

So if I wanted to drink milk for calcium, fat-free is my best bet, but if I'm drinking milk for fortified vitamins, full-fat is?

Yes — if choosing a food to prioritize some of its nutrients over others really makes sense. These nutrients are important, but there are many other ways to get enough of both: For one thing, many other foods are fortified with one or both of these nutrients now. For another, we make vitamin A from precursor nutrients like beta-carotene (but also many other carotenoids) in plants, and we make vitamin D when our skin is exposed to sunlight. For that matter, you could skip the dairy and get both nutrients in supplement form, too. (See page 184 for more on vitamin D and on nutrient supplements.) And regarding these added nutrients, the differences between whole milk and skim milk are very modest — not a reason, in our view, to choose one over the other. Do that because of the saturated fat content and calories, which vary a lot, not because of calcium and vitamins A and D, which vary much less.

Just tell me, is cheese good for me or should I stay away?
The best answer is that you should eat it sparingly, if at all: Cheese is a concentrated source of calories that you probably don't need, it's a concentrated source of saturated fat that probably isn't beneficial and may be harmful, and it's a concentrated source of sodium, which most of us eat too much of to begin with. You certainly don't *need* to eat cheese.

Highly processed cheese concoctions like "American" cheese — usually labeled "cheese foods," the dominant source of calories in many cheese-based junk foods — aren't really even cheese. They have added coloring agents, flavoring agents, and so forth, and all the dairy that goes into these products is manipulated in ways we can't imagine.

Earlier you said that the fat in cheese and yogurt might be different than that in milk. What do you mean?
Fermented dairy has a long list of interesting saturated fatty acids with differing metabolic effects. Cheeses come from different animal sources and are produced with a variety of cultures, leading to unusual combinations of fatty acids with varying health effects. Right now, no summary judgment is possible; there may be benefits — or problems — that we don't yet know about.

How about other fermented dairy, like yogurt?
You definitely want to avoid sweetened yogurt — the sugar and other additives completely change the equation. High-quality plain yogurt has the benefits of milk along with some probiotics, which are good. And the fermentation process literally changes the fat compounds of the dairy; like cheese,

there are novel saturated fatty acids in yogurt that aren't as common as other saturated fats.

How is my microbiome affected by fermented dairy?

Cheese and yogurt are fermented with active cultures, and those active cultures seem to have positive effects on the microbiome (see page 189). Needless to say, this is an area of evolving research. Our bottom line is this: Since there's no established health benefit of fermented dairy, eat it only in moderation, and only if you like it.

There are many protein-rich foods where a health benefit is reliably established — beans, lentils, seeds, nuts, and whole grains to name a few. If you're trying to eat to optimize health, you don't need to add cheese or even yogurt to that list. This is not to say you should avoid cheese and yogurt if you like them, just that you shouldn't eat them as staples and you shouldn't eat them because you think they'll save you.

MEAT

A big one: Should I eat meat?

The prevailing idea, one that's easily countered, is that we need to eat meat to be big and strong. But horses are a lot bigger and stronger than we are, and they don't eat meat; neither do elephants, the biggest, strongest land animals on the planet.

The fact is that *animals can make muscle out of whatever food they are adapted to eat.* Humans are omnivores, adapted to eat both plant and animal foods. We can make all the muscle we need out of either: We have choices. Meat is optional.

But isn't meat the best protein source?

The single best selling point for meat would be the delivery of high-quality protein, but here's the thing: As we discuss on page 151, we don't need as much protein as we've been told.

We keep looking for "one size fits all" answers, but in the real world, the answer far more often is: It depends!

Most of us get *too much*, and we can get all we need without meat. And let's be clear: A good source of protein and a good protein source are not necessarily the same thing. A "good" source of protein that harms both your health and the environment could deliver a concentrated dose of high-quality protein. But if a food that is good for your health and much better for the environment delivers ample protein with all the quality you need, then, we believe, that is a *better* source of protein. See beans, page 96.

Is meat healthy in a strictly nutritional context?

There's an essay, written by experts at Curtin University in Australia, titled "Meat is a complex health issue but a simple climate one: the world needs to eat less of it." That pretty much captures it.

The health argument is subtle: If you mostly eat plants, there's almost certainly room for meat in your diet without clear evidence of adverse health effects. But if your diet is excessive in saturated fat from many sources (lots of full-fat dairy and processed meat), meat isn't good for your health. And if you don't eat a lot of plants, meat is one of the first things to replace with them.

This much is clear: Meat is either absent entirely, or eaten in small amounts, in all the valid contenders for "best diet" for human health.

Does the treatment of an animal determine how "healthy" it is to eat?

Of course the quality of the meat is an important consideration. You see a significant shift in the composition of the meat depending on the dietary and activity pattern of cattle (for example), which points to the advantages of meat that is pasture-raised and grass-fed; compared to grain-fed beef, it's lower in total saturated fat and higher in omega-3s. There's a bigger difference still between minimally processed meat and hot dogs.

But it's important to note that these are still small differences when compared to the nutritional differences between meat and beans.

If I'm a vegetarian, should I consider adding well-raised meat to my diet?

There's no evidence of any kind suggesting that the addition of meat to your diet produces a health benefit. Regardless of the quality of the meat, beef replacing beans is a downgrade.

Well, is meat an important food for anyone to include in their diet?

If you're living in a place where there is ample food available, then the answer is no. Virtually no one in the US suffers from protein deficiency, and no one "needs" to eat meat.

The exception is when protein deficiency occurs, and that's mostly due either to famine — actual starvation — or, sometimes, in homebound elderly people. There is a condition called sarcopenia, which is an age-related loss of muscle; meat is helpful in preventing that.

By the way, this sort of thing explains why we hear all the time about conflicting diet studies! We keep looking for "one size fits all" answers, but in the real world, the answer far more often is: It depends! Starving? Meat is much better for you than starvation. Overfed in America? You don't need meat, and would be better off without it. Clearly, both of those can be true, and are. But you can see how, when used to produce titillation rather than illumination, they could be the basis of competing headlines.

Why have we been led to believe that meat is the best source of protein?

One of the reasons there's so much fascination with meat as a protein source has to do with the established professional definition of the biological quality of protein (which we think needs to change). Another reason, of course, is the lobbies whose interests lay in promoting animal products.

How is protein defined now, and how does this definition need to change?

The current definition is put together only by considering the protein concentration in a serving of food (like 20 grams per serving), by the distribution of the essential amino acids present, and by the digestibility of those amino acids (essentially, how readily proteins in the food can be broken down and brought into the bloodstream).

What's missing from this definition of "protein quality" is the quality of the food delivering the protein. The problem is that we are using the word "quality" in a way that's misleading. It refers only to the protein in a food, not to the food delivering the protein. But the latter matters far more, especially since, as we've noted (see page 151), you do NOT need to depend on any one food for your protein; you can get it from any balanced variety of foods over the course of a typical day and week. And you already do. If you are reading this book, we guarantee that you are not protein-deficient. Food quality should reflect what eating that food does to our health in general, and maybe the health of the planet, too. The biochemical definition of "protein quality" does not consider these matters at all, and may very well direct people away from them.

If you say something is a high-quality protein source, that makes it sound like it's something I want.

Exactly. When, in fact, beef is not preferable to other protein sources. Remember: We want to center our diets around food, not the nutrient properties that make up food. Here's a good analogy: Imagine a coal-burning train delivering solar panels. Solar panels = good for the environment; coal-burning trains = not so much! Yes, you need protein, but there are better sources than meat.

Let's ask, "Does the addition of this food to the typical diet improve the diet and health outcomes?" And also ask, "Does the consumption of this food confer harm or benefit to the environment at large?" The best, modern definition of "protein quality" would reflect not just the quality of protein in a food, but more important, the net quality — in all ways that matter — of the food delivering the protein. When this is done, beans and lentils — not beef and pork and poultry — top the charts.

That sounds like a much more holistic approach.
It really is. We also think it better reflects how we all use the word "quality," which is to mean something preferable, something we want. For something to be a high-quality protein source, first and foremost it should be a food that you recommend people eat more of to be healthier. It's incongruous that the "highest quality" protein sources are otherwise discouraged for people who want to improve their health. Misleading, in fact.

Regardless of protein, I like meat!
That's a good reason to eat it; pleasure is one of the ways we choose what to eat. But you probably can cut your intake: As a group, Americans should be eating 90 percent less meat than we do.

So it's okay if I eat it in moderation?
Meat in moderation is actually a big step toward opening up room in your diet to get more nutrients from plant foods and

to consume less saturated fat. The less meat you eat, the more room there is in your overall diet for, say, beans or lentils. The only trouble with "moderation" is that it is pretty vague, and it may or may not be a good threshold, depending on where you are drawing the line.

But leaving the environment aside, what's wrong with meat?
Beef, especially, is a concentrated source of saturated fat, of which we eat too much already. High levels of saturated fat elevate "bad" cholesterol (LDL) in the blood, elevate inflammation, and clog up our arteries. All these consequences increase our risk of heart attack, stroke, diabetes, cancer, and dementia.

Does it matter which part of an animal I'm eating?
To some extent, the cut you're eating is an important consideration. That they're different in composition is obvious just by looking at them. Meat that has obvious fat, like ribs or bacon, has more saturated fat than lean cuts. But lean meat is problematic also, so this isn't the biggest issue.

And let's be clear: Like so many arguments about nutrition these days, the arguments we hear about saturated fat are sometimes misleading and unhelpful. The key issue is not whether saturated fat, per se, is good or bad for us. Imbalance is bad for us, always; good diets are balanced diets. That's a universal truth.

You also said beef harms the environment.
It's hard to talk about beef and not talk about environmental impact, because the environmental footprint of beef is pretty much off the charts relative to everything else. It's associated with a massive contribution to greenhouse gas emissions; replacing beef with beans, for example, would reduce our greenhouse gas emissions by something like 50 percent. And since

there are no healthy people on a cooked planet, this really does need to be part — arguably, right now, the most important part — of the conversation about health.

Why is it so bad?
Cattle produce methane in their GI tract, and they release this methane into the atmosphere, and methane is a singularly potent source of atmospheric carbon. And, of course, cattle consume enormous amounts of grain and water; they're the least efficient protein source there is.

If the global population limited beef consumption to famine or protein deficiency, it would mean an enormous reduction in the environmental impact of food production, more equity and social justice in the food system, and better health all around.

What if I switch out my hamburgers for a grilled chicken sandwich?
Again, it's an "Instead of what?" situation. Yes, chicken is better than beef, though it's also a concentrated source of saturated fat, akin to lean cuts of beef. (Turkey is somewhat leaner.) But even then, what the birds were fed and whether or not they got any exercise makes a difference. And the fact that you're eating meat probably means you're not eating beans or other plant foods, which is always the better choice. (And by the way, the white bread that sandwich is on is likely to be nutritionally worthless, or worse than that.)

How about free-range poultry?
Free-ranging animals are leaner and generally healthier to eat than confined animals; animals that eat a native diet (for chickens, this means scraps and grubs as well as grain; for cattle, it means grass) are also better for you. All animals are what

they eat, so the composition of their muscles is influenced by their diet and, of course, the exercise they get.

What about pork?

Pork is marginally preferable to beef, though nutritionally it can be a cut-by-cut thing. Pigs come in second to beef in greenhouse gas emissions, though, and at this point almost all pig farming is done on large industrial farms, which produce a great deal of pollution and, of course, concern themselves little or not at all with animal welfare.

And lamb?

Lamb is somewhat leaner than beef (though this too varies with the cut and lifestyle of the animal), and sheep are way more likely than cattle to be fed their native diet and raised in circumstances where they routinely exercise. It's important to know what the animals are being fed and how they are being raised and treated. That's why sourcing meat locally is ideal: The best way to know that is to see it happening on a farm that is somewhere near you. (See page 79 for more about eating locally.)

How about bison and wild animals?

All meat choices are not as "high quality" sources of protein (by our new definition on page 115) than alternative plant sources would be. Against a backdrop of diets that are excessive in saturated fat already and deficient in a lot of plant-based nutrients, the conclusion about all sources of meat should be the same: They should make up a relatively small portion of the diet.

Having said that, wild animals are in one way the best choice, if for no other reason than that it's the one you're likely to make only on occasion. If you said you ate meat, but only when you could get wild meat — or really, really well-raised meat — you'd be talking sense. Of course, 8 billion hungry *Homo sapiens* can't say this, and mean it, unless we all eat less meat, period.

FAKE MEAT

What's your advice about meat alternatives?

Well, for starters — what alternatives, and what meat? The net benefits of substituting plant foods for meat/animal foods depends on the quality and composition of both. It's more important, for instance, to displace processed meat from your diet than to displace unprocessed meat; more important to displace fatty, factory-farm-raised meat than lean, locally sourced, pasture-raised meat; more important to displace beef than fish; and so on. And then there is a whole range of plant alternatives that also vary in composition and benefits.

I was thinking about "fake meat" like Beyond Meat and Impossible burgers . . .

You and everybody else these days. The particular advantage in these products, which certainly qualify as "ultraprocessed," is that, when it comes to the population at large, they can potentially win over people who really like to eat meat, eat it routinely, and are *not* inclined to give up beef for beans, or those people who'd like to eat less meat but have trouble doing so. Whether these products offer a health benefit is uncertain, but they certainly have lower environmental impact than meat, and are obviously better for animals — two good reasons to consider them. Whether you're an omnivore or a vegetarian, they can be good options to satisfy the occasional burger craving, or a convenient meat-free option if you're out and about or at a restaurant. As usual, the question is "Instead of what?"

Instead of a conventional Whopper? Sure. But instead of almost any meal cooked from scratch using basic ingredients? Probably not. This is ultraprocessed food.

Simply put: If you're inclined to eat meat, make it a small part of your diet, and make it as pure as possible — from ethically treated, well-fed, well-tended, well-exercised animals. When you're not eating meat, the more you stick with whole plant foods, on their own or as ingredients, the better; they're better for us, our fellow creatures, and the environment.

So what's the bottom line?

What meat you are replacing, and what you are replacing it with — both matter. For the sake of the planet, the less meat — especially beef and other mammals — we all eat, the better. If you're unwilling to forgo meat unless your veggies imitate it, then by all means try Beyond, Impossible, and related offerings that tempt you. But if you're willing to stick with whole plant foods on their own or as ingredients — that's even better.

FISH

Fish is healthy, right?

Eating fish is clearly and decisively good for people when the baseline diet is anything like the prevailing American diet. Again, remember, "Instead of what?" Fish instead of meat: for sure. Fish instead of plants: maybe not. But unlike terrestrial animals, fish (at least fatty fish) are a significant source of omega-3 fatty acids (see page 131).

Can I get too much omega-3?

We don't know. The Inuit live mostly on marine animals and have the highest intake of omega-3 of any group. But whereas

omega-3 is beneficial to those of us who tend to get too little, it's not clear that the Inuit benefit from getting that much.
They don't have especially high rates of coronary disease, but
they don't live a particularly long life, either. So there seem to
be limits.

What about the toxins in fish, the ones that keep pregnant women away from spicy tuna rolls?

There are concerns about contaminants in fish, particularly in
predatory fish like swordfish and big tuna. Bioaccumulation
is the process by which toxins (like mercury) in small fish are
concentrated in their flesh, and then further concentrated in
the larger fish who eat the small fish, and so on. That's why the
highest concentrations of water-borne toxins — whether it's
heavy metals or chlorines — occur in large fish.

Why are these heavy metals and other poisons of serious concern when you're pregnant?

You've got a very small fetus undergoing rapid development,
and concentrated exposure to heavy metals from fish (or any
other source) would be problematic: You're exposing the
adult body and that much smaller body to the same dose of
toxins. There are also special vulnerabilities while a body is
forming during embryology that fully-developed bodies don't
have.

Are larger fish still something for non-pregnant people to avoid?

Yes — or at least, to limit. But even eating those large predatory fish is associated with better health outcomes, despite the
bioaccumulation of toxins, compared to diets without fish. In
other words, the net benefit of eating fish outweighs the harm
of the toxins, as long as you're not pregnant. It makes sense

that you might want to mix up the type of fish you eat, so don't eat large tuna or swordfish all the time, for example.

What's the healthiest type of fish?

That would be wild-caught fatty fish like salmon, halibut, and mackerel, fish that are rich in omega-3. Fish that are low in omega-3 are still a good source of lean protein, generally way better than land animals.

So if I ate an optimal vegan diet, would it be beneficial to add fish?

We can't say; the research has never been done, and there's no way to figure it out with available data. The ideal comparison would last a lifetime, to compare the effects of, say, an optimal vegan diet and an optimal pescatarian diet on longevity and lifelong vitality. But short-term studies that look at a variety of biomarkers would be a good place to start. These can and should be done, but they haven't been.

The overall evidence we have suggests that adding fish to conventional, modern, Western diets, where it will likely displace other kinds of meat, is beneficial. But there is nothing to suggest that adding fish to diets where it would displace whole plant foods confers any benefit.

How about the environmental impact of eating fish?

The impact of fishing is that we're destroying the oceanic eco-system, devastating whole populations, making species extinct, even wrecking the ocean floor and upsetting ecologies. And much aquaculture (fish farming) is similar to land agriculture — it creates pollutants, uses high levels of resources, and features some despicable labor issues. From an environmental perspective, it makes sense to limit fish consumption. Again, should fish replace other animals in your diet? Yes. Should it replace plants? No.

So: What's your final recommendation on fish?

For the sake of the oceans and the planet in general, we need to limit our intake and focus on sustainable sourcing, maybe on invasive species. For health, we should eat smaller fish more often: herring, mackerel, sardines, anchovies, butterfish — these are highly concentrated sources of omega-3 and least in danger of overfishing; and, because they're closer to the bottom of the food chain, less subject to bioaccumulation of toxins.

Mollusks, like mussels, clams, and oysters, have been farmed for thousands of years with little negative impact. And wild fish from healthy populations, like Alaska salmon, are also pretty safe to eat.

Generally, think about sustainability and be conscious of the environmental impact of eating fish. Whatever the net effects of eating fish on human health — and those are positive — one thing is very clear: There aren't a lot of fish left. So, like other animal products, eat fish only occasionally.

COOKING OILS

Okay, we've talked about a lot of food, but not about what you cook food in. Tell me about the different types of cooking oils.

This is an area where there's a lot of folklore and passion — heated, um, past the smoke point.

The facts: Oils are collections of fat molecules; they all include a variety of fatty acids, each of which is a particular kind of fat molecule. Those fatty acids occur in families: saturated fat, monounsaturated fat, and polyunsaturated fat — and then there are subdivisions of those families. There are no oils that are completely free of saturated fat or unsaturated fat, and none that's all omega-6s or omega-3s. They're all combinations.

Does it matter what oil I use? If I'm going to be making an effort to cook spinach and quinoa, I don't want to cancel out their health benefits.

You're not alone: Many people are concerned about which oil to use. And the answer is, "Yes, it matters." Unfortunately, it's not simple, and even more unfortunately, this topic — like almost everything else in nutrition — has been overtaken by ideology in place of science.

What do you mean?

The pop culture argument in recent years made the case that coconut oil was a superfood, little short of a panacea, and that canola oil was somehow responsible for crimes against humanity. Both of these notions are false.

Then there are the people who say that all cooking oil is a processed food, and therefore the best amount to add to the diet is none. ("It's the sugar of fats," we've been told.) To us, this seems an ideological argument, considering that some of the healthiest, longest-lived populations on the planet have diets that derive considerable numbers of calories from oil, particularly from extra-virgin olive oil. (And by the way, extra-virgin olive oil is barely processed.) And there's another camp, the Paleo camp, if you will, that argues that like dairy, oil isn't part of humans' original diet.

People in the Stone Age didn't use cooking oil, so neither should we?

Basically. It's true that humans only relatively recently invented mills to grind seeds or other foods into oil; but on the other hand, fat dripped out of meats and fish as they cooked; that fat was undoubtedly collected and subsequently used in cooking other foods. We think it's presumptuous to obsess over the exact details of when oil extracted from foods was first introduced into human eating. We really don't know. And since we aren't experts, we've checked the work of those who are — the paleoanthropologists — and they really don't know either.

What about the health basis of the "don't add oil" argument?

The most famous, early studies of whole-food, plant-based diets to treat and reverse heart disease occurred in the late 1970s, 1980s, and early 1990s, when the prevailing view was that we should limit total dietary fat intake. The two perspectives of plant-based (i.e., vegan) and low-fat were blended in the resulting "intervention" diet. We really think this was something of historical happenstance, but it is what it is.

These studies showed regression of coronary atherosclerosis (that is, arterial plaque buildup shrunk partly away) specifically, and no other diet has done that, possibly because no other diet has been studied that particular way. But other diets, like the high-fat Mediterranean diet, have been shown to lower heart disease rates just as much, and to us, that's what really matters.

It's true that we don't have a study that shows that a diet rich in extra-virgin olive oil shrinks coronary plaque, but that's not because it's been shown *not* to do that, but because the study simply hasn't been done (absence of evidence). The population-level data suggest that when the fat is good fat, well-balanced, plant-predominant diets, whether high-fat or low-fat, are comparable in benefits.

But it's better to use oils than animal fat, yeah?

For the most part, yes. But there's an interesting historical sequence here. You go back far enough, and animal fats, like lard and beef tallow, prevailed in the US. Then there was the understanding (in part ushered in by biologist Ancel Keys and the Seven Countries Study, which began in the 1950s) that switching from animal fats to more plant fats was associated with plummeting rates of heart disease. People realized that plant fats were better.

Were there any benefits to using animal fats for cooking?
Animal fats are more highly saturated, which means that they're more stable, with a longer shelf life and the ability to hold up well to heat. They're good for preserving and cooking; they're not so good for health. (We have a rule of thumb: The longer the shelf life of a food, the shorter the shelf life of the person eating the food! There are exceptions, of course — beans and whole grains keep forever — but it's worth bearing in mind.)

The longer the shelf life of a food, the shorter the shelf life of the person eating the food!

But longer shelf life is a major benefit for the food industry. So when studies about the health damage of animal fats came out, the industry desperately searched for equally versatile plant fats. They settled on palm oil, palm kernel oil, and coconut oil. And they developed trans fats.

Huh. I thought coconut oil was a recent "discovery."
Coconut oil has recently become the darling of superfood fantasies, but it's long been a substitute for animal fats, because it mimics their culinary and production properties. So do the other tropical oils, like palm oil and palm kernel oil. These became the principal oils used in food service and food production. But it was a sideways move, because the tropical oils contain high amounts of saturated fat. We were swapping one source of saturated fat for another.

Then what happened?
Things got even worse when food engineers took mostly unsaturated cooking oils from plant sources and bombarded

them with hydrogen. That process, hydrogenation, twists many of the molecules to produce a configuration known as "trans." Producers had altered the natural state of plant oils into something more like beef fat.

And it worked out really well for them. The fats weren't fully saturated but partially hydrogenated. Trans fats have those same desirable properties as saturated fats: long shelf life, stability, and high heat tolerance. But they have even worse biological effects than saturated fats. They are horrible for our bodies.

Why?

These compounds are highly inflammatory and contribute to the propagation of atherosclerosis (arterial plaque buildup) more than anything else. This move away from saturated animal fats to hydrogenated vegetable fats — trans fats — was an unmitigated public health disaster from more than one perspective.

Ultimately, the FDA took action against trans fat, and in 2018, it was banned from the US food supply; other countries around the world banned it even before we did.

That's promising!

Yes, but we didn't need to be in that position in the first place. And there was another terrible result from the trans fat debacle.

That was the "you can't trust nutrition experts" effect. Because nutrition experts were largely responsible for the transition from tropical oils to trans fats. It's likely that at least part of the reason that coconut oil came back with such passion attached to it was the "see I told you so" idea — that we listened to nutrition experts and nothing good came of it. The other explanation for tropical oils being partly back in vogue is that the food service industry needed to replace trans fats with something, and tropical oils are the best "something" available for many applications.

So what's wrong with tropical oils?
Tropical oils aren't terrific, but they're better than trans fats. Still, they're higher in saturated fat than oils that are liquid at room temperature (tropical oils are solid at room temperature). And production of palm oil — the dominant oil used in many countries — increases at the expense of the rain forests, which, as you probably know, not only encourage biodiversity, but are massive carbon sinks, great protectors against global warming.

And coconut oil?
As with so many stories in processed food, that of coconut oil — especially its recent resurgence — begins with a surplus. And the best way to sell surplus coconut was to turn it into oil. So coconut oil began to be marketed as the superfood oil.

It's not? It won't help brain function and weight loss?
Coconut oil is not a superfood. (As we discuss on page 138, neither is anything else.) Coconut oil won't fix your life or your bad diet. Neither will hemp, or flax, or anything else. However, although coconut oil is high in saturated fat, the predominant fatty acid it contains is lauric acid, a relatively short-chain saturated fat that behaves differently from the saturated fat in meat and dairy.

How is lauric acid distinctive?
Some inconclusive literature suggests that lauric acid lacks the tendency to promote inflammation and atherosclerosis. There may be evidence that the predominant fat in coconut oil lacks

the overt harms of other saturated fats, but there's no indication anywhere in the scientific world that it offers health benefits. "Lack of likely harm" is a pretty low bar to set for a good food, let alone one with pretensions of being "super." *There are no proven health benefits associated with coconut oil.*

But I thought coconut oil was this natural, whole-food alternative to more processed cooking oils.
That's a different argument — that coconut oil is good for you because other oils are processed with chemicals and altered in the production. "Less bad" is not exactly "good," though it may be "better" or "preferable." And coconut oil can be processed just like other oils; there are better and worse types of coconut oil.

This is what the refined and unrefined varieties of coconut oil are about?
Exactly. All vegetable oils can be extracted from their sources with mechanical forces or with chemical processes. With coconut oil as with other oils, you can extract the oils simply with pressure, in a cold setting. That's usually called cold-pressing, and it's preferable: The structure of the oil isn't changed in that process.

Alternatively, you can use a sequence of steps that expose the oil to heat and a variety of chemicals (including hexane, a petroleum derivative), and here you both strip nutrients from and alter the composition of the fatty acids, possibly making the oil a bad actor. All vegetable oils — coconut, canola, olive, soybean, sunflower, corn, and so on — may be cold-pressed or chemically extracted using solvents. And, of course, they may be organic — or not.

So in general . . . ?
There is a pretty reliable benefit to just shifting from animal

fats to plant fats. And no matter what oil you use, you want it to be cold-pressed, if possible, and organic, if possible.

So choosing organic cold-pressed coconut oil ensures I'm getting a healthy cooking oil?

No. It ensures you're getting the best coconut oil you can. But organic cold-pressed coconut oil is still oil that's high in saturated fat and doesn't provide omega-3s; there's no evidence it provides any health benefit at all. It's not as bad as some oils, but not as good as others.

I'm kinda down on coconut oil now.

The main reason to include coconut oil in your diet would be because you like its culinary properties, not for any health benefit. Just like butter or lard. In fact, lard has a more beneficial fat profile than coconut oil — with more unsaturated fats and less saturated fat. That's not to say you should start eating lard, but that any real food can be included in small amounts in an otherwise healthy diet; no food should be eaten to excess; and there are no superfoods.

Some types of oil are better than others, right?

Yes. All oils that are liquid at room temperature are healthier than those that are solid at room temperature. That's not much of a refinement, since most are liquid. A more challenging and legitimate controversy is whether we should be concerned about omega-6 fat. There are prominent nutrition experts who say, "As long as we shift from animal fats to plant fats, and as long as we shift from saturated fats to unsaturated fats, all will be well." (Or, at least, a heck of a lot better.) There are others who raise concerns about the imbalance that tips toward omega-6 polyunsaturated fat and away from omega-3.

I've definitely heard about omega-6s and omega-3s; what's the difference?

Long-chain omega-3s are important for brain development, eye development, metabolism, the production of hormones, and checking inflammation. Although omega-6 is a class of essential fatty acids, Americans tend to get too much of it, and at high levels it becomes pro-inflammatory, and can interfere with the body's ability to manufacture omega-3.

And we get an excess of omega-6 from cooking oil?

Most oils used in processed foods include something on their label like "May contain one or more of the following." Inevitably that list includes soybean, corn, cottonseed, and/or sunflower oil. Many of those are labeled "cooking" or "vegetable" oil, too. And they're all rich in omega-6 fat and low in omega-3.

But omega-6 is an unsaturated fat, and aren't unsaturated fats preferable to saturated animal fats?

Yes, but this subdivision is important: Remember that omega-6 tends to promote inflammation; there's concern that an excess of omega-6 is not good for us. It's not that omega-6 fats are "bad" in and of themselves — they're essential nutrients. But we're getting too much of them. Our diets are out of balance.

Is there a specific ratio of omega-6 to omega-3 that I should be looking for in oil?

Although no scientific study has yet found it for sure, we have guidance from the work of paleoanthropologists trying to piece together our native, Stone Age dietary intake from fossil records and other sources. Such experts estimate a range of omega-3 to omega-6 of about 1:1 to 1:4. This suggests that we should all be getting at least a quarter as much omega-3 as omega-6, and maybe equal amounts daily. But today's typical American diet yields as much as twenty times omega-6 as omega-3. That's a big distortion.

That's bad?
There's debate about whether what matters most is the absolute amount of omega-3 or the ratio of omega-3 to omega-6. But in the absence of definitive science on the topic, and in looking at our native diet, we know we get too much omega-6 and too little omega-3. That may matter less than the balance between unsaturated and saturated fats, and/or between plant fats and animal fats. But there's reason to think it does matter.

We like the general rule: Balance is good, imbalance is bad. (That's true for more than diet.) So, if diets are providing an excess of omega-6 and a relative deficiency of omega-3 in general, then more omega-6 exacerbates an existing imbalance, and, on general principles, we vote "no" to that.

Which oils have a high ratio of omega-3 to omega-6?
Flax, hemp, walnut, olive, and avocado oils all have very high concentrations of omega-3. And, of course, fish oil, though you can't use that for cooking or salads.

Any oils with a high omega-6 content that I should avoid?
Corn, safflower, and sunflower oils (and, formerly, soybean oil — stand by) are all high in omega-6. However, that doesn't mean you have to avoid these altogether.

What does it mean, then? I'm confused.
These high-omega-6 oils are present in almost every packaged food you consume, so if you want to shift the balance of fatty acids in a favorable direction, the best way to do that is to cut back on packaged food. That's a good thing for everything we're talking about in terms of building a better diet.

Then, when you're home, you'd be better using one of the oils that's high in omega-3. These tend also to be high in mono-unsaturated fat (MUFA), which is not pro-inflammatory and, in general, is preferable; olive oil is the best-known of the oils high in MUFA. But — and this will surprise you — canola oil is good, too.

What? I've been avoiding canola oil for years thinking it's harmful to my health.

Canola oil comes from a plant called rapeseed. Before the rapeseed plant was selectively bred, it contained something called erucic acid, which is one of the more toxic fatty acids. But decades ago, selective breeding cultivated varieties of the rapeseed plant to remove all or most of the erucic acid. (The word "canola" is a bastardization of "Canada oil," made from a variety of rapeseed developed in Canada specifically designed to produce oil from these low–erucic acid varieties.)

Now canola oil is a great choice for health. In fact, a switch from animal fats to canola oil was a particular focus of the famous North Karelia Project in Finland, which began in the 1970s and is associated with a more than 80 percent reduction in rates of heart disease and an addition of ten years on average to life expectancy.

Canola oil has good fat?

Canola oil has a stunningly good fatty acid profile. It's overwhelmingly unsaturated fat, which is a good thing. In addition, it's a fairly rich source of the same kind of monounsaturated fat found in olive oil. It's very low in omega-6 fat, and there are certain varieties of canola (and usually it will say this right on the bottle) that are significant sources of omega-3 fat.

Canola oil also has an extremely mild flavor — it's what we call a "neutral" oil — so it's quite versatile for cooking. It also stands up fairly well to heat. There's almost no downside.

All of this makes it a really good daily oil; maybe even the best there is. Since there aren't whole populations that have been eating canola oil–rich diets for many generations, we don't have the evidence base for the benefits of canola oil that we have for olive oil. But its fatty acid profile makes it a contender.

How is canola oil processed?
Like every oil, you can make it by mechanically pressing the seed (cold-pressing) or by chemically extracting the oil. The better choice is always cold-pressed (and organic).

Isn't canola oil made with GMOs?
We talk about GMOs with regard to soybeans on page 98 — the same argument applies here. Except for this: Rapeseed hasn't been genetically modified! The rapeseed plant used to produce canola oil was selectively bred, not genetically modified. So saying "non-GMO canola oil" is like saying "gluten-free eggs." (Don't put either claim past the marketers.)

Are there any caveats?
Just two, and the first we've already said: Buy organic cold-pressed canola oil. (Do that with other oils, too.) And if you buy more than you're going to use in a week or two, store the bulk of it in the refrigerator so it doesn't go rancid. (This, also, is true of all oils. If it's liquid at room temperature, it can spoil at room temperature.)

I always thought olive oil was the healthiest oil.
It probably is. When we compare cooking oils for the greatest and clearest evidence of benefit, cold-pressed extra-virgin olive oil wins, hands down. And when you consider that it's also among the best-testing, whether you're cooking with it or and eating it raw ... there's really no contest. Some other oils —

canola, walnut, avocado — may be as good constitutionally. But none has olive oil's track record.

What's the proof that olive oil is the best option, health-wise?

We have evidence from intervention studies that show cardio-vascular benefit in healthy adults; from studies showing reduced heart attacks in adults with heart disease; and from mechanistic studies that look at its effects on insulin, glucose, lipids, etc., all of which are positive. There are, as well, detailed studies of components of extra-virgin olive oil, in particular a bioflavonoid antioxidant called oleocanthal, showing beneficial effects on everything from platelet stickiness to cell mutation rates.

> We know that generation after generation, a diet rich in olive oil produces lots of people who live long and enjoy vitality.

But mainly, and most important, we have this: Some of the world's longest-living and healthiest populations are in the Mediterranean region, where olive oil is a staple, and has been for generations. We know that generation after generation, a diet rich in olive oil produces lots of people who live long and enjoy vitality. We can't say that about canola or any other oil.

Olives themselves are highly concentrated sources of antioxidants, and some of the antioxidants found in olives are not found in any other food. Oleocanthal is under intense investigation, and it seems to have favorable effects on mutation rates and cell culture studies, on endothelial function (good endothelial function means good blood flow, with potential for improved immune system function, cardiovascular health, memory, sexual function — anything that benefits from blood flow, basically), blood lipids (fats, like cholesterol), inflammatory markers, and more.

What makes people think that olive oil is the key to their longevity? Couldn't it be . . . rosemary? Or some other Mediterranean ingredient? Capers?
We're not sure anyone thinks that, exactly. Remember how many times we've said, in one way or another, to eat whole-some foods in sensible combinations? Olive oil isn't a super-food. But it is among the key components (neither rosemary nor capers qualify, much as we like them) of the famously healthful Mediterranean diet. That diet has been studied, no-tably by the "godmother" of the Mediterranean diet, Professor Antonia Trichopoulou of the University of Athens. (See page 35 for more about the Mediterranean diet.)

What does it mean for olive oil to be "extra-virgin"? That's the best kind, right?
Olive oil, like all other oils, encompasses a whole range of products depending on production methods. There are a lot of different production methods, and the labeling is different than it is for other oils.

It's pretty simple: Extra-virgin olive oil is cold-pressed; there are some other EU standards that it must meet, but they're arcane and only relevant in that they mean, generally speaking, that it's "good." "Virgin" olive oil is also cold-pressed, but meets less exacting standards. (You don't see it much.)

"Pure" olive oil (which isn't pure) may be a combination of cold-pressed ("unrefined") and refined (chemically extracted) oils. (You may also see "pomace" olive oil; definitely refined.) "Light" olive oils are all chemically extracted. So extra-virgin and virgin are cold-pressed; they may or may not be organic; that is, the olives may or may not have been sprayed with chemicals.

Do all types of olive oil contain those antioxidants you were talking about, like oleocanthal?

Production method does matter. If you use chemicals or enzymes to extract the oil, if that oil sits around or is exposed to light for a length of time, the fatty acids can be completely degraded and the antioxidants lost. So the olive oil no longer has any of the properties intrinsic to the nutritional composition of the olive. In fact, if the oil has been abused, it's bad — and bad oil of any kind can impair health.

Extra-virgin olive oil provides by far the most concentrated dose of antioxidants — a good indication that tinkering with ingredients as little as possible provides the best results. In addition, extra-virgin olive oil gives you a rich amount of a monounsaturated fat called oleic acid.

And monounsaturated fat is good for me?

Yes. Monounsaturated fat is associated with a range of health benefits. It tends to preserve or bump up levels of so-called good cholesterol, or HDL, which is associated with lower heart disease risk; while helping to lower levels of so-called bad cholesterol, or LDL, which is associated with increased cardiovascular risk. It's associated with favorable effects on blood sugar and blood insulin, which means it can reduce inflammatory effects.

Are there arguments against olive oil?

A small but vocal group of health experts claim that olive oil is bad for us because there are studies that show it impairs endothelial function rather than aids it. But that information comes from a small, fifteen-year-old study that didn't mention what olive oil they were using. So for all we know, it could have sat in the sun for several days and then gotten run over by a bus. Certainly there was no indication that it was extra-virgin olive oil. In any case, we think you can discount this. The evidence for the benefits of using olive oil is overwhelming.

What about the other popular oils, like soybean oil?
Soybean oil, like canola oil, is a product of the decades-old effort to move away from saturated fats. Again, the big issue is
in how it's processed. Almost all soybean oil in this country is
chemically extracted.

Does that mean I should avoid it as a cooking oil?
It means, as with other oils, you should look for cold-pressed
(and preferably organic) oils. In any case, most soybean oil has
a fat profile that's between those of canola oil and olive oil. It's
slightly higher than olive oil in monounsaturated fats, so a very
good source of oleic acid. And it's extremely low in omega-6,
and a good source of omega-3.

All these factors combine to produce a sound all-purpose
oil, as long as it's cold-pressed. By the way, until fairly recently,
commodity soybean oil was quite high in omega-6 fat, and
omega-3 was low or absent. But new cultivars of soybeans
are producing an oil with just about an ideal fatty acid distribution, very comparable to omega-3–rich canola oil. But as
of this writing, this variety isn't widely available. For now we
think you're better off sticking with canola oil.

SUPERFOODS

**Is salmon a "superfood"? What about goji berries, spirulina,
or açai?**
You don't need any particular food to have a great diet. If you
have a great diet, you will derive maximum health, vitality, and
longevity benefits from it, and no so-called superfood will add
to that. No single food can suddenly make you a healthy person; it's the overall dietary pattern that matters.

So how does something become a "superfood"?
What tends to get labeled a "superfood" is something with

interesting nutrient propositions, especially foods that concentrate antioxidants, like pomegranate, green tea, and dark
chocolate. The second property of a superfood is that it's often
considered exotic, food we're not eating routinely. The whole
thing is a marketing ploy, and a transparent one, that feeds industry and sensationalist media.

**Another defining characteristic of superfoods is that they
are expensive!**
Oh yeah. They're expensive because they're superfoods! They
sound like miracle drugs, which is what so many of us are looking for. When a food is being introduced, hypermarketed for
the first time, and it's a novelty, and you can claim a high nutrient property for it . . . that's a superfood.

But all these foods are really good for us, correct?
Sure. They're great. But no better than a million others. Is a
pomegranate better than an apple or an orange? Debatable.
Against berries of all kinds, more debatable still.

So why are you against the name "superfood"?
The implication is, "This is a food that can do what only super
things can do. It has superpowers." And that's simply not true.

**Let me guess: Instead, we should be focusing our energy on
sustainable, healthy dietary patterns?**
Yes, please! Overwhelmingly, the evidence links large-scale dietary pattern, not the addition of "superfoods," to health. One
of the reasons we struggle so in our apparent confusion about
diet is that there's a lot of research attempting to isolate the effects of any one food. That's nearly impossible to do.

So it's all bunk?
This doesn't mean that some foods aren't better than others;

just that they're not panaceas. Dark chocolate is a good example: In moderation, it's shown to have positive health outcomes, particularly in cardiovascular health.

So I can eat a bar of dark chocolate a day?
Well, just because something is healthy doesn't make it okay to eat in large quantities. Especially something that's likely to be at least 30 percent sugar.

What's the sweet spot of how much dark chocolate to eat, then?
We don't know. Nobody has answered that question. If we were to make an educated guess, we would say something like an ounce of dark chocolate that's 70% cocoa or higher a day is probably okay in a typical diet.

So the only reason I need to watch how much dark chocolate I'm eating is because of the sugar?
You got it. If you routinely eat sweeter sweets, like donuts, you could make an upward move to dark chocolate. At the end of the day, an ounce of dark chocolate daily is probably good for health — plus, it's a source of pleasure.

Like red wine!
Very much like red wine. See page 147 for more on that.

DRINKS

So now I have a better idea of what to eat to be healthy, but what should I drink?
Let's start with what humans have drunk historically. First, mother's milk, then water.

Our ancestors would occasionally crack open a coconut or squeeze a fruit and drink what came out of there. But we have pretty clear indications that they were high-endurance athletes compared to us, so they needed hydration much more than we do, and water got the job done.

Then why hydrate with something other than water now?
For most of us, there's no reason to. Maybe if you play high-intensity sports — because you're losing electrolytes in your sweat. If you're hydrating with water while you're using a lot of sodium and you don't replace the sodium, you could conceivably start to have low sodium levels in the blood. People once took salt pills in those circumstances; now they drink "sports drinks." Not the worst thing in the world, but only in the middle of a high-intensity workout are they justifiable. (The best option to regain those electrolytes would be to eat, but you don't want to do that in the middle of a workout.)

Can I still benefit from the hydrating effects of Gatorade if I'm not a high-level athlete?
Nope. Watching football is not the same as playing it.

Okay, I get that I don't need it, but is drinking Gatorade going to be detrimental to my health?
Probably, yes. If you're not working out intensely, you're getting sugar and calories you don't need from sports drinks; the liabilities way outweigh the benefits.

What should we be drinking?
Water.

What type of water, exactly? There are those various infusions in water, like alkaline?
Don't be fooled by all the so-called enhancements to water. It's true that the modern diet tends to shift more toward acidic, so a lot of the foods that are best for us are more alkaline. But you

don't have to alkalize your diet with "enhanced" water; better if you just eat more of the food — leafy greens, beans, fruits — that tends to help alkalize the body naturally.

Enough about water—let's talk about coffee. Is it possible that a cup of joe not only benefits my productivity at work, but also benefits my health?
It is! Recent studies suggest that routine consumption of coffee is associated with long-term net health benefit.

Well, that's awesome news. What type of benefits?
The antioxidants in coffee are shown to improve cardiovascular health, and of course there's the short-term benefit of a caffeine boost. There are studies that show caffeine enhances performance on cognitive tests and improves concentration. So if you get the dose right — if you're not jittery on caffeine, and it's not sabotaging your sleep — routine incorporation of coffee into your diet is beneficial. Then there's the fact that if you're drinking more coffee, you're probably drinking less soda.

Frappuccinos and lattes included?
Decidedly not. And you knew that already. Those drinks are sugar bombs filled with chemicals, which totally changes the equation; the coffee is an innocent bystander in this case. We're talking about coffee in its native state, with nothing added to it.

Black coffee has a pretty dense concentration of antioxidants. Because the typical American diet includes a lot of high-calorie foods that are nutrient dilute, coffee actually ranks right near the top as a source of antioxidants. (If we ate berries, brightly colored vegetables, or dark chocolate every day, it would knock coffee down the list.)

Does the way the coffee is brewed affect its antioxidant levels at all?

There is a difference between filter (or drip) coffee and French press coffee. There are some fat-soluble compounds in coffee, called antimetabolites, that potentially bump up cholesterol levels and have an inflammatory effect. They are filtered out quite reliably in filter coffee, but they are present in French press coffee. That's something most people probably don't need to think about, but it might be clinically meaningful if you drink French press coffee exclusively, especially if you drink multiple cups a day.

For the best health, should I get my daily dose of caffeine from coffee or tea?

Although tea is a more concentrated source of antioxidants than coffee, we can't fully answer that question because there's never been a long-term study comparing coffee drinkers and tea drinkers. While the antioxidants in tea have the potential to defend against a variety of chronic diseases — including cancer — coffee is more narrowly linked to cardiovascular benefit. So it's impossible to say that one is better than the other; they're both good choices.

Is green tea healthy?

The routine consumption of green tea, which is rich in antioxidants, has been suggested to do all sorts of wonderful things for us; at least some of them are true. The evidence isn't all that strong, which leads us to a conversation about science and evidence. Imagine how difficult it would be to isolate just the effects of green tea on health. You could look at people who routinely drink green tea out in the real world, but they don't just drink green tea in a vacuum; they have dietary patterns and lifestyle patterns associated with drinking green tea.

So if you notice that populations that routinely drink green tea have less chronic disease than a population that chronically drinks soda, you could say the green tea is good . . .

. . . or you could say the soda's bad.
Right. Or you could say we're not sure because there are a lot of other things that differ between these two populations.

What type of ideal study would you design to figure out whether green tea has a positive effect on overall health?
We would need to randomly assign people to drink either green tea or something else. The question immediately becomes: What is the something else? Is it green tea versus water? Green tea versus milk? Green tea versus beer? Green tea versus vodka? Green tea versus coffee? And then whatever differences you see might be because of the green tea or because of the other replacement beverage. Or because of both.

So like questions of which food is best, it really does depend on what green tea is replacing. Pretty unsatisfying non-answer, I've got to say.
Sorry, but that's the way nutrition science is. Marketers may want you to believe otherwise so that you buy their products, but really it's not only about any single food or drink, but what it's replacing. If tea replaces soda, it's great. If it replaces coffee, it's probably a wash. If it replaces water . . . it's probably okay, as long as the caffeine doesn't bother you, and assuming you don't doctor it with cream or sugar.

How about black tea?
The difference is in the fermentation. Green tea and black tea come from the same plant, but black tea leaves have been fermented, while green tea leaves have not. Black tea is lower in antioxidants than green tea, but it's still good. In terms

of its health benefits, it's the same deal: It depends what it's replacing.

And white tea?
Though it's rarer than green and black teas, white tea is made from the leaves of the same plant; for white tea, the leaves are picked before they unfurl. White tea actually has the highest concentration of antioxidants. But it's not easy to find, and it's expensive.

I'm surprised it's not a superfood.
You're right. As much as anything else, it should be. It's worth pointing out that while you can say that black and green teas are loaded with antioxidants, infusions — teas brewed from herbs or spices — resist generalizations. You can make flavored hot water with just about anything.

Winding down instead of up, is alcohol actually good for me?
For a long time, the prevailing wisdom in medicine has been that alcohol is the quintessential double-edged sword. There seems to be a narrow therapeutic window, meaning there's a dose that's probably good for you. But at the same time, if you go above that dose range by even a little bit, the harm accumulates quickly.

Can you spell out the harms when you go past the threshold of that therapeutic window?
The harms are in two basic buckets. There are the acute harms of alcohol because of what it does to our judgment and our cognition. Then there are the longitudinal harms because of what it does to our organ system.

I don't need a lecture about drunk driving. What does alcohol do to our organs?

It really does pickle the organs — in particular, the liver. Excess intake of alcohol over time leads to scarring of the liver and ultimately cirrhosis, or liver failure, which is truly horrible. It also damages the brain.

How much alcohol is too much?
It varies by individual. We are all dependent on an enzyme called alcohol dehydrogenase to metabolize alcohol, so how much of that enzyme you have determines how much alcohol you can tolerate. That varies by sex and by ethnicity. Men, in general, have twice as much alcohol dehydrogenase as women.

Why?
No one knows. We just know that men, on average, can reasonably process twice as much alcohol as women.

Alcohol tolerance isn't just a weight thing?
No. Even matched for size, men can tolerate more alcohol than women. And then there are significant ethnic variations. Famously a lot of ethnic Asians have a very low tolerance for alcohol, which simply means they have low levels of alcohol dehydrogenase.

So how much alcohol are we talking if I want to stay within the therapeutic threshold?
The prevailing average is one drink a day for women, two drinks a day for men. A "drink" is essentially one mixed drink, one beer, or one glass of wine.

How do we know this?
There have been a number of studies that have looked at a single population with various rates of disease and compared different levels of alcohol intake. The lowest rates of cardiovascu-

lar disease in particular were seen in people who drank in that narrow therapeutic range.

So we know this? It's been established?
Yeah, it's pretty valid.

Can I save up my one or two drinks per day and drink the same cumulative amount on the weekend?
Binge drinking causes what's known as a cardiomyopathy, or holiday heart syndrome, which impairs heart muscle function. So, no. If you want to stay in the therapeutic threshold, you can't just have seven drinks on Friday night.

Okay, okay. What are the benefits of drinking alcohol moderately?
Alcohol, or to be specific, the ethanol in alcohol, bumps up levels of protective HDL cholesterol pretty effectively, actually as effectively as intense exercise. The other thing that ethanol does is make platelets a little less sticky.

Why does that matter?
Because platelets sticking together is ultimately the final step in the chain of events leading to a heart attack.

So there's a cardiovascular benefit.
Yes, and there is even more of a cardiovascular benefit because of the antioxidant content in alcohol.

Right, antioxidants are important! Are they only in red wine, though, or are they also in IPAs or martinis?
There are some antioxidants in most alcoholic beverages, including the gin and vermouth used to make a martini, and in beer. Red wine is generally considered the best alcoholic beverage for antioxidant content, however.

Why red wine and not white wine?

While red wine and white wine come from the same grapes, red wine is pressed with the skins, where the antioxidants in grapes are mostly found, and white wine is pressed without the skins.

It happens that some of the populations with famously long life and good health routinely consume red wine.

So I can rely on drinking red wine moderately to reduce my risk of heart attack?

Let's not get carried away; the effect is pretty modest. Rather, ask this: "Can I drink wine because I like it and, oh, by the way, it's potentially good for me, too?" That's the right idea; don't think you can rely on alcohol as your primary means of avoiding a heart attack.

At the population level, health outcomes are best among people who drink no alcohol at all. Research on that topic translated into headlines that basically said we have been wrong all along, and that any amount of alcohol is harmful, because the level associated with the least net harm in any population is zero. Bad effects of alcohol range from bad judgment to pancreatitis and cirrhosis to car crashes.

So this study revealed that it's better to just swear off alcohol than to drink one or two drinks a day?

Yeah, but there's a problem with the logic in this study, or at least in how it was translated into news. In a population, any level of drinking could be achieved in a variety of ways. You could have a population of 200 people where everybody has one drink a week. Or you could have 199 people who have zero drinks and one person out of that whole group who has 200 drinks a week. So, the safest alcohol level for a whole population is zero, because that's the only way to be sure alcohol won't harm anyone. But that is very different from research to see what level of alcohol intake is associated with the best health outcome for any given individual.

So who would actually benefit from adding, say, a glass of red wine to their diet, and who is better off just staying away?

If you eat well, exercise, don't smoke, and, oh, by the way, drink a moderate amount of alcohol, will we necessarily be able to detect the effect of a glass of wine a night on your cardiovascular risk? No. You probably won't have a heart attack whether or not you had the glass of wine. Right? Because you are at low risk to begin with. The opposite is true as well: If you smoke, eat badly, don't exercise, and, oh, by the way, also have a glass of wine, it will not prevent the harm from all those other things.

If you drink and get into a car crash and kill yourself, we're gonna notice that. So at a population level, the harms of alcohol are very visible. They tend to be acute and obvious, while potential benefit to those members of the population drinking responsibly and moderately are going to be hard to capture.

Is there a best-case scenario where having a drink or two a day actually provides meaningful health benefits?

Here it is: You are a man or a woman with a family history of cardiovascular disease. You're taking pretty good care of yourself otherwise. You like wine with dinner. You understand that it's one to two drinks a day (one if you're a woman) and no more. You don't drink before you drive.

You stay in that narrow therapeutic window. It's probable that you would derive some benefit. But we wouldn't say drink just for that reason, especially if you didn't like to drink already. You can see it's iffy.

> Pleasure is good for health! We tend to be stressed, and relieving that stress with pleasure contributes a health benefit in its own right.

What's your final recommendation re: alcohol?
The sensible advice about alcohol would be: If you drink it because you like it and carefully stay within the therapeutic range we've discussed, yes, you may derive some discernible health benefit, especially when it comes to reducing the rate of cardiovascular disease. And pleasure is good for health! We tend to be stressed, and relieving that stress with pleasure contributes a health benefit in its own right. Just be careful about where you seek that pleasure, because some sources of pleasure are rather ill-advised. But if you don't drink, there's no good health reason to start.

Nutrition 101: Macronutrients, Micronutrients, and Body Responses

I wish there was a magic formula, like, all I have to do is get a certain combination of nutrients in a day and I'm guaranteed good health.
There is: It's called eating the right foods; it's called "balance." We've already discussed that. But if you want to talk nutrients, we can. You know about recommended daily allowances?

Yeah, I've heard of them. Where do they come from?
They originate with the National Academy of Medicine (which is part of the National Academy of Sciences). Based on evidence they collect from intervention studies, growth in children, and the aging process, they establish target ranges for both macronutrients — protein, fat, and carbohydrate — and micronutrients, like vitamins, minerals, trace elements, and things like carotenoids (a type of antioxidant).

PROTEIN

How much protein do I need?

Almost certainly less than you think. The recommended range for protein is the narrowest of all macronutrients, but still pretty broad. Recommendations say we should be getting from 10 to 35 percent of our calories from protein; that would mean about 50 to 175 grams daily. Average daily intake among adults in the US now is above 90 grams, with men eating 100 grams on average.

What do people mean by "complete" versus "incomplete" protein?

The difference between complete protein and incomplete protein is their essential amino acid content. Any time you hear the word "essential" referring to a nutrient, it means it's something the body needs but can't make: You have to get it from food (or a supplement). Sodium is an essential nutrient; so are certain amino acids. Essential amino acids are protein building blocks that the body needs to function properly, and we can't make them. A "complete" protein source is one that provides *all* of the essential amino acids. Nearly all protein-containing foods do that, although some have more essential amino acids than others. Contrary to what you may have learned in high school, in general it's not something you need to worry about.

But isn't it true that only animal products have complete protein, have all the essential amino acids? Isn't it true that if you are vegan, you'll have trouble getting enough protein?

NO. All essential amino acids occur in plant foods, although they occur in varying concentrations. "Incomplete" protein means that that food doesn't contain *all* the essential amino acids. But *most* foods, including plant foods, actually do.

Our understanding needs an update: What used to be

thought of as incomplete protein might be better described as "less concentrated" protein, with lower levels of specific amino acids. That's one reason it's important to eat a variety of foods: You'll get ideal amounts of all essential amino acids. But the body's needs can be met readily and certainly with a balanced, varied diet of exclusively plant foods.

So if I'm not eating meat, I should try to eat all the right plant foods to cover all the essential amino acids?

You're working too hard: If you have a balanced vegan or vegetarian diet, you'll be getting ample amounts of all of these essential amino acids. You don't need to make a special effort to eat broccoli rabe or lingonberries! The "balance" we keep referring to is not based on some mystical formula known only to biochemists; it's really just a variety of wholesome foods.

The distribution of essential amino acids in different plant foods — nuts and grains and legumes and vegetables and fungi and seeds — is famously complementary; what you don't get from one, you'll get from another. You don't need to worry about getting them all in every meal.

The signature benefit of meat and eggs is that they provide essential amino acids in ideal proportions all at once. But many variations on the theme of optimal eating, including a quality vegan diet, can deliver the ideal balance of protein over the course of a full day, and that's what matters.

So protein—amino acids—can come from beans or beef or protein shakes?

Spot on. Your body really is agnostic as to where those amino acids come from. Since we are omnivores, if we have all the essential amino acids required to build muscle, we can make muscle out of anything — plants or animals. Then what really determines the muscle bulk is exercise.

I'll just take it day by day, then?

You don't even have to worry about food combinations in a given day. You really only have to worry over the course of any stretch of days.

The body is capable of banking amino acids for up to 72 hours. Think of this like materials being dropped off at a construction site: If you have the lumber but not the hardware, the lumber will sit there until the hardware shows up. The key thing is to get everything there in time for essential construction. Much the same is true for our bodies: If you eat a decent diet of good food, it'll all get there in time; you actually don't have to worry at all.

So no one needs to worry about getting enough protein, ever?

No one who is food secure. If you get enough food, you'd have to work very hard to be protein deficient, with an unbalanced, quirky diet. Other than that, you will never wind up protein deficient. It's all but unheard of in the US, or any part of the world where people aren't starving.

If I want to bulk up in muscle, can I go over the 35 percent recommendation?

It likely depends on how much "bulk" you have in mind. If you are a competitive bodybuilder, there is some evidence that very high protein intake may help your cause. But otherwise, no: Put in the work and you will put on the muscle. Despite popular belief, more protein doesn't make you bigger and stronger; exercise does. Extra protein is helpful for that competitive muscle mass, but otherwise, you won't benefit from it. It's not even healthy.

It's not even healthy???

Protein intake above 35 percent stresses the liver and kidneys and is taxing on the skeleton. So, after you win Ms./Mr. Universe, dial it down!

You're saying too much protein could be detrimental to my health? I've only ever been worried about getting enough.
Too much of most nutrients, including protein, can be detrimental. Protein is acidic, and too much of it can be bad for your bones. Also, extra protein generally means extra calories, and extra calories — whether they come from protein, fat, or carbohydrates — turn into body fat. That, you don't want.

The idea that we all need more protein than we get and the more, the better — is pure pop culture folklore. Very good for those selling miscellaneous protein powders or making the case that you need to eat buffalo to have big muscles. But not so good for you!

I guess I have this idea ingrained in my head that big, strong guys eat a lot of meat.
The simple fact is that we can convert food into muscle, and that includes plant food. A horse takes in oats, grass, and hay and turns it into horse. A gorilla takes in leaves and grasses and turns it into gorilla. Horses and gorillas are "big, strong guys."

What should bodybuilders and athletes be eating, if not a bunch of protein?
Are you sitting down? Carbohydrates.

No way. Like bread?
"Carbohydrates" does not mean only bread. All plant-based foods are carbohydrates. (All foods contain all three macronutrients, but plant-based foods are heaviest in carbohydrates.) Studies show that diets that foster high performance and speedy recovery generally favor complex carbohydrates over

protein. You obviously need enough protein (and fat, too). But as we've said, you're already getting that.

CARBOHYDRATES

What are carbohydrates, exactly?

Just to get this framed reasonably right from the start, plant foods are carbohydrate foods. "Carbohydrate" refers to the molecular structure that makes up most plants: a chain of carbons with a water molecule attached to each, or literally, "hydrated carbons." Hence ... carbohydrate. All of them, from broccoli to pasta to sugar. The structure of plants is largely made up of carbohydrate, just like the structural components of mammals are mostly protein and fat. (And water, of course. Everything living is mostly water.) Humans have got carbohydrate in the mix, too, but a lot less of it. Plants are kind of our mirror image; they are mostly carbohydrate. What *kind* of carbohydrate matters.

> Carbohydrates can be anything from lentils to lollipops, pinto beans to jelly beans.

Wow, give me a second to process. Carbs aren't only in Wonder Bread, but are actually also in kale?

Told you to sit down. But yes! Carbohydrates are the largest source of nutrient energy on the planet, by far, and carbohydrate food is plant food. Most of the food energy that feeds all the animals on the planet comes from carbohydrates.

Then why does everyone think carbs make you gain weight?

Because carbohydrates come to us not only in the form of plants in nature, but in many forms made in factories. Carbohydrates can be anything from lentils to lollipops, pinto beans to jelly beans. A summary judgment about such a vast swath of food is utterly absurd. It's like saying, "Weather is dangerous."

Some carbs are the staff of life; some are the stuff of disease. *Cutting all carbs would mean cutting the most nutritious foods from your diet.* Oddly, a number of popular diets ask you to do just that: For example, the Whole30 diet wants people to cut out beans and whole grains. This is a huge mistake, both for health and weight control.

But if by "cutting carbs," you mean "stop eating candy, junk food, dessert, and white bread, and stop drinking soda" . . . well, great. That's the right move. Just remember that "carbs" means "plants," too, and you want more of those, not less.

Carbs can be "good" or "bad," then.
Right. Think of apples, spinach, and carrots versus Apple Jacks, cinnamon rolls, and cotton candy. They're all "carbs." If a diet cuts out relatively unprocessed plant carbohydrates — fruits, vegetables, and whole grains — it's fallen for the classic peril of lumping too much together. You can't dislike or renounce carbohydrates — they make up the most nutritious foods on the planet.

If carbohydrates are as vast as the weather, which ones are more like a blizzard than a warm spring day?
Refined grains, like white flour, are a big one. When manufacturers strip out all the good stuff, you are left with a flour that is low in fiber and closer, as far as your body knows, to sugar than to a whole grain. And, of course, so is added sugar, under all of its many aliases: sucrose; high-fructose corn syrup; brown and white; invert and cane; honey and agave syrup; organic brown rice syrup; and all its other guises — you get the idea.

When the bulk of your carbohydrate consumption is refined flour, stripped of its fiber, plus added sugar, carbohydrates become a bad actor.

So carbohydrates CAN be the bad guy.

It's not really bad carbohydrates but bad *foods*. You can make bad foods out of anything. You can make bad foods out of carbohydrate, you can make bad foods out of protein, you can make bad foods out of fat.

Trying to capture that with generalizations about nutrients, like "protein good, carb bad," is either misguided or a marketing ploy.

I get that foods made with highly processed grains like white bread and donuts are unhealthy. What about whole grains?

Whole grains are good for us. They are one of the best sources of fiber we can process. Large-scale studies consistently find a relation between routine whole-grain consumption and lower risk of all chronic disease and cardiovascular disease, lower blood pressure and blood lipid levels, and improved glycemic control (the fluctuations in blood sugar levels).

What is fiber, and what does it do for us?

Fiber is a special kind of carbohydrate that the body cannot convert to sugar, and it does many good things: It mitigates the potential adverse effects of concentrated carbohydrates, like high glucose or insulin levels. And it's very good for the GI system, too; it helps move things through the system quickly, which is good.

Fiber comes in two main varieties: soluble and insoluble (sometimes called viscous and nonviscous, respectively). Soluble fiber dissolves in the water of the GI tract and creates a filter or barrier there, slowing the entry of nutrients such as sugar and fat into the bloodstream. Insoluble fiber does not dissolve in water, and thus represents bulk in the GI tract, which stimulates transit through the intestines. Neither gets into the bloodstream — both exert their effects in the GI tract, where they can also serve to nourish the microbiome.

And I can only get these types of fiber from unprocessed carbohydrates?
Right. Good sources are fruit, veggies, beans, lentils, berries, and oats and other grains. (Some grains are good sources of only insoluble fiber, some of soluble fiber, and some of both.) If you strip out the fiber but preserve the starch — that's what happens when you make white flour — you totally change the net health effect of eating that food. You bump the glycemic load (we'll get to that on page 172) into the stratosphere. We must learn to trust the wisdom of nature and leave well enough alone.

What about the claim that carbohydrates stimulate insulin release?
They do; so does protein, and the greatest release of insulin happens when we combine protein and carbohydrates. But it's important to remember that the same plant sources that are concentrated delivery vehicles for carbohydrates also give us all our fiber, and soluble fiber smooths out the glucose and insulin response curves, preventing dangerous spikes. This is an important difference between carbs with fiber and carbs without — which are inevitably ultraprocessed: "Good" carbs require insulin for metabolism, but also dial down the amount of insulin needed. High intake of soluble fiber actually *lowers* blood insulin and blood sugar levels.

Is there a reason to be conscious of my overall carb intake, even when it's coming from whole grains, veggies, and fruit?
No. The Atkins/keto argument is wrong: As long as carbs, fat, and protein are in balance, and they're the right kind — few or no ultraprocessed carbs, little or no saturated fat (and no

trans fat), and few or no animal products — then the diet is a good one.

Carbohydrate, plant-based foods are most closely associated with more vitality, less chronic disease, lower rates of heart disease, and lower risk of diabetes and obesity. Those foods are vegetables, fruits, beans, lentils, and whole grains, and one thing they all have in common is that they are rich sources of carbohydrate, including fiber. So, no, don't worry about total carbohydrate at all; worry about the kind of carbohydrate. Focus on wholesome foods in any sensible combination and let the macronutrients sort themselves out.

> An apple isn't a "carb"; it's an apple.

So how many carbs should I really be eating?

Remember: Carbs are plants. Plants are carbs. And you can get not only carbs but protein and fat from plants. But okay, to answer that question directly: Carbs should make up the majority of your diet, no matter what style you eat; all the world's healthiest diets focus on carbs. The Food and Nutrition Board recommends a carbohydrate intake beginning around 40 percent of your daily calorie intake and going up to as high as 65 to 70 percent. If you're eating carbohydrates in a healthy way, this includes a wide array of whole foods.

And one more thing: The idea that carbohydrates make us overeat is not true. It's much easier to overeat cheese or ice cream than apples; overeating depends on the specific food. Again, it's the whole food that determines the net effect on health, not some part of the food. An apple isn't a "carb"; it's an apple.

So "carbohydrate" is too broad a term to actually tell us something about how we should be eating?

Right. We have these ranges for macronutrients, but really you could just eat whole foods and not worry about any of this. It's nice to understand it, but it's pretty simple to remember the apple/Apple Jacks thing. No?

FAT

How much fat is acceptable to eat in a day?
Despite everything we "learned" in the late twentieth century, the range for fat is considerably broader than other macronutrients. There are claims for good outcomes with diets that have 10 percent or less of calories from fat (like in Okinawa); those would be the low-fat diets. There are Mediterranean diets that get well over 40 percent of their calories from fat and seem to produce the same great health outcomes.

Of course the kind of fat matters, and we'll get to that.

That's a pretty wide range!
Yeah, it is. But within that range, if the rest of the diet is sound, it'll produce good health outcomes, as long as the fats are predominantly polyunsaturated and monounsaturated. These are the fats found naturally in nuts, seeds, and fruits like olives and avocados, which are vastly different from the fats found in steak and cannoli.

But isn't fat the macronutrient most associated with weight gain?
It's the macronutrient with the most calories per gram: nine, versus four for protein and carbohydrate. That's a big deal: Each gram of fat has twice as many calories as one gram of the other two macronutrients. And ultimately, calories are what matter most to weight. So the same weight of fat will make you fat compared to protein and carbs. But fat can make you feel full faster, depending on the source, resulting in you eating less.

Twice as fast? Sometimes. Some very high-fat foods, such

as nuts, are associated with better control of appetite and weight. But in a high-fat diet that's also high in processed foods and sugar, the energy density of fat becomes a major liability. This is why there is a lot of science suggesting that high-fat foods can contribute to obesity, but some literature pointing the other way. In this, as in everything else, it depends on the food sources; it's about the food, not the nutrient.

An apple has 100 calories; so does an ounce of cheese, or about four dice-size cubes. Think about eating five apples; think about eating twenty little cubes of cheese. Most people are definitely more likely to do the latter. Foods like cheese and ice cream are easy to overeat. A cup of ice cream — not a huge serving for most people — is around 300 calories. But ice cream is a source of both fat and sugar (as is pizza, the crust of which contains hidden sugar). That combination is especially problematic, because fat is the most concentrated source of calories, and sugar the most potent stimulator of appetite. Beware that combination, ye who want your pants to fit!

Can we talk about saturated fat? Some people are now saying that saturated fat is actually good for us.
It's a false narrative, but before getting into that, remember that it's balance that matters. Saturated fat, per se, isn't bad for people; too much is. Since our diets tend to provide all the saturated fat we need, and much more, additional saturated fat is certainly "bad" for us. Yes, still.

The new "sat fat is good" argument is partly based on the idea that during the low-fat craze in the 1990s, we got fatter and sicker. Ergo, cutting fat was a bad idea.

But we never really cut fat! Fat intake in the US has mostly gone up over recent decades, but since total calorie intake has gone up *even more*, with much of that from refined carbohydrates and added sugar, fat as a percent of total calories has gone down a bit. We got there not by shrinking the fat/calories numerator — but by growing the overall calorie denominator!

As for the idea that saturated fat, per se, has been exonerated, that all started with two meta-analysis research papers that focused on telling us what we need to know about saturated fat. Both observed ranges of saturated fat intake across various populations, from relatively lower to relatively high, and found that there were no differences in rates of heart disease. So their conclusions were that saturated fat can't be the bad actor we thought it was.

That's surprising, no? Well, these studies overlooked a vital piece of information: The rates of heart disease were high and constant both with a higher and "lower" consumption of saturated fat. For starters, that means that both dietary patterns — whatever they were, exactly — were comparably bad for the heart. The "lower" levels were still pretty high, and still above the recommended level for saturated fat intake. Everyone included in these two papers had too much saturated fat in their diets — it just ranged from a little too much to a lot too much. The findings were at best oversimplified in the media — and perhaps it's fairer to say they were just flat-out misrepresented.

But there's an even more important flaw, in that the studies didn't ask this question: For the people eating a little less saturated fat, what replaced it? Neither of those papers addressed what people were eating *instead* of saturated fat. In fact, the word "sugar" doesn't appear even once in the more recent of the two papers (published in 2014). Yet when people cut fat, they have tended to turn to things like high-sugar snacks, many of which contained trans fats — the "SnackWell's phenomenon." In other words, people who ate a little less saturated fat may have been eating both more sugar and more trans fats. You would think the authors would have explored that

"Instead of what?" issue, because there's more than one way to eat badly.

Has there ever been a study that looks into what replaces foods that are high in saturated fat?

Yes. In 2015, a study that followed more than one hundred thousand people over thirty years asked this question: Among the group who reduced their intake of saturated fat over time, what did they replace those calories with? They found that some people reduced the percent of calories from saturated fat — so a bit less dairy, a bit less meat — and increased their intake of trans fat; presumably, they swapped out butter for margarine. And in that group, the rates of heart disease went from bad to *worse*. So: Yes — you can do worse than eat saturated fat. That's why trans fat has been consigned to the dustbin of historically bad nutritional ideas. (It's been banned in the US.) It also demonstrates how you can call saturated fat "good." But just because a broken leg is better than a heart attack . . .

What foods should I cut out if I want to reduce my saturated fat intake?

Meat and dairy. And replace those saturated fat calories with unsaturated fat calories or complex carbohydrate calories. That's a huge win, both ways. When sat fat calories are replaced with either whole-grain calories or unsaturated fat calories, rates of heart disease are markedly lower.

What foods with unsaturated fats should I be eating?

Nuts, seeds, olives, olive oil, avocados, fish, and seafood.

But all those foods are still high in fat. Wouldn't it be better to switch to low-fat options?

Not necessarily. Remember that people who replaced saturated fat calories with a low-fat option typically ate foods with re-fined grains and sugar instead. That would mean they ate less pepperoni pizza, but more SnackWell's cookies. Their rates of heart disease were exactly the same. The type of fats and the quality of the food sources seem to matter much more than any given total amount of fat in the diet. There is plenty of healthy fat out there, as we'll get to.

Are there any good saturated fats?
Yes: The saturated fat that predominates in dark chocolate — stearic acid — isn't associated with inflammation or athero-sclerosis. There's also debate about lauric acid, the saturated fat that predominates in coconut. Different saturated fatty ac-ids behave somewhat differently, and there are saturated fatty acids in cheese and yogurt that may prove innocuous, or per-haps even beneficial. But to be clear, there's no convincing ev-idence of any benefit from any dairy fat yet — as there clearly is for, say, the fat in extra-virgin olive oil, or walnuts, or wild salmon.

What's the takeaway on saturated fat?
There is more than one way to eat badly, and thanks to the food industry, we have been encouraged to explore them all! Nothing in the mix says that saturated fat is good for us, but that doesn't mean that trans fats, ultraprocessed grains (like white flour), or sugar are good substitutes. High levels of all of these lead to high levels of heart disease.

CHOLESTEROL

Should I be worried about cholesterol?
Yes, if you're worried about heart disease. Elevating levels of cholesterol (a fat-like compound in animal foods, and also something the body makes; it circulates in the blood and is

used in building a variety of cellular structures and hormones) in the blood — especially "bad" cholesterol, or LDL — increases the propagation of atherosclerotic plaque, the stuff that gums up our arteries, causing heart attacks, strokes, and some kinds of organ failure. LDL stands for "low-density lipoprotein," and of the several compounds that transport cholesterol through the bloodstream, this is the one most closely linked to elevated heart disease risk. In contrast, HDL, or high-density lipoprotein, is often referred to as "good" cholesterol, because it's the cholesterol transporter that best helps clear cholesterol from the bloodstream — and it is associated with reduced risk of heart disease. That there is a link between cholesterol levels in the blood and cardiovascular risk — the higher your cholesterol levels, the higher your risk of cardiovascular disease — is known as the cholesterol hypothesis, and that hypothesis is almost universally embraced by experts in heart health.

If it's a hypothesis, does that mean we're not sure about cholesterol?

No. We've known for seventy-five years that you do not want high levels of blood cholesterol. There is no debate in mainstream cardiology over this. There is a very small but highly vocal contingent who argue that cholesterol isn't a bad actor and that we don't have to worry about our intake of saturated fat or dietary cholesterol. They are mostly the same voices that would argue for eating more meat and more dairy. They're wrong.

What's the difference between blood cholesterol and dietary cholesterol?

Blood cholesterol is what's in our bloodstream; dietary cholesterol is what we eat. For many reasons, dietary cholesterol has a lower impact on the body than blood cholesterol. And dietary cholesterol actually has a lesser effect on blood cholesterol than many other things that we eat. It has a much lesser effect than saturated fat, for example. In fact, our bodies have a pretty good system to get rid of excess cholesterol.

What foods contain cholesterol?

Meat, dairy, and the most famously concentrated source, eggs. Cholesterol is present to one degree or another in almost every animal food, including fish and seafood, since it is a part of animal cells.

Are there foods that lower blood cholesterol levels?

Many of the usual healthy suspects: soluble fiber, like that from beans, lentils, certain grains, and a wide variety of fruits, berries, and apples. Monounsaturated and polyunsaturated fats can't solve anything definitively, but they can help because they usually replace the consumption of saturated fat.

What even incited the debate over cholesterol?

The controversy heated up with the release of the 2015 Dietary Guidelines Advisory Committee Report, which is the scientific work that informs the development of the American dietary guidelines. Scientists decided that the typical American diet isn't contributing enough cholesterol for it to be a significant concern. Therefore, having people focus on it is no longer helpful. What they didn't say is that dietary cholesterol is largely irrelevant.

But we should instead focus on saturated fat intake.

Right. That affects blood cholesterol levels way more, anyway.

Dietary cholesterol is still bad for us, though, right?

Advocates of very healthy diets, like an optimized vegan diet, say you shouldn't consume dietary cholesterol, and there are studies that show that if you are eating a high-quality vegan diet and add an egg a day, you will see a negative effect on blood cholesterol. Essentially, if you start with almost no exposure to dietary cholesterol at all — and vegans would have none, because dietary cholesterol is only in animal products — and then you add dietary cholesterol, it will bump up your blood cholesterol levels.

In contrast, against the background of a typical American diet, which is quite high in saturated fat and has a fair amount of cholesterol, there really is no clear association with small variations in dietary cholesterol.

So what you're saying is that we have bigger fish to fry than worrying about cholesterol in eggs.
Yes, exactly. There's no real public health value to having people focus on how much cholesterol they eat. And there's an interesting tangent here: Think "America runs on Dunkin.'"

How do donuts fit into this conversation about cholesterol in eggs?
When Americans stop eating eggs because they're concerned about their cardiovascular health, they have not usually switched to oats, walnuts, and mixed berries, but instead to donuts, bagels, or a Danish. By focusing on cholesterol and telling people not to eat eggs, we're doing net damage to the overall quality of people's diet because of what has typically replaced eggs.

The question, again, is, "Instead of what?" Eggs are certainly good for you if the alternative is a glazed donut. They're nutritious and satiating, and their benefits outweigh the potential harm from their cholesterol. The upgrade from eggs are the usual whole grains, fruits, vegetables, and so on.

So if I'm eating healthy vegan breakfasts, I shouldn't add eggs; but if I eat donuts and the like, I should embrace them?

Yes, both of those ideas are completely valid. Unless you're vegan, you can have eggs in your diet. As a protein source, they're a great way to avoid other sources of cholesterol, like saturated fat. So if you are having more eggs, have less meat, which has at least twice as much saturated fat.

But really this is a discussion of lack of harm. If we want to talk about *benefit*, the better breakfast is nuts, berries, and oats — for example.

INFLAMMATION

What exactly is inflammation and why is it a bad thing?
Inflammatory responses are the native action of the immune system. And the important thing about the immune system is that it represents the body's basic defense against foreign invaders. Its fundamental role is to differentiate self from non-self. So in fact, inflammation is usually a *good* thing.

Is anything from outside my body considered a non-self invader?
Yes, and that can be a problem. The immune system presumes the worst of the non-self. It treats anything that gets inside the gates as a threat.

That sounds okay.
Much of the time, it is. When all is said and done, we are pretty well protected against a world full of bacteria, viruses, parasites, and fungi because the immune system defends us with inflammation. Inflammation occurs anywhere in the body where there is a perceived threat. The perceived threat triggers an influx of white blood cells and chemicals, which we call inflammation. Sometimes you can even see it: If an invader gets in and a part of the body gets infected — let's say you get

a splinter — that part of you turns red and swells up. It's in-flamed.

So inflammation allows us to literally see the body healing itself?

The inflammation we can see also represents inflammation we cannot see. It's the same basic process, but it occurs in the in-ner nooks and crannies of the body, like our blood vessels and inside our vital organs.

So to be clear, we actually need inflammation?

If we didn't have inflammation, we wouldn't heal wounds. We wouldn't be able to effectively fight viruses, parasites, and fungi — or the rogue cells that cause cancer. So, yes, we need an inflammatory response.

But it's not always what you want?

Unfortunately, it can be difficult for the body to distin-guish between self and non-self; they can wind up looking a lot alike. And there can be bad consequences in both direc-tions.

How so?

When non-self looks too much like self, it can sneak in and elude the immune system. Then, when self looks too much like non-self, it contributes to autoimmune diseases — such as rheumatoid arthritis, Crohn's disease, or multiple sclerosis. Or it can make us prone to things like allergies to tree pollen. Tree pollen is not dangerous in itself, but because the immune system recognizes it as a non-self invader, it can act as though pollen is a threat. The result is a whole lot of misery that isn't really protecting us from anything. So far as we know, no hu-man has ever died of . . . being pollinated.

So why is inflammation itself bad?

The inner battle of self versus non-self has the same potential consequences as military conflict: Innocent bystanders — in this case, the body's healthy cells — can get hurt. In the case of inflammation, you're waging battle inside the body, releasing chemical compounds designed to kill bacteria and viruses. Ideally, inflammation avoids that by striking the optimal balance between a robust response and a careful one. Many of those compounds use oxidation to fight invaders. As rust does to iron, our white blood cells use oxygen free radicals to kill enemies within. But they can kill healthy cells, too.

This is the reason why balance in inflammatory responses is crucial. If you have too little, you are ill defended against actual threats. If you have too much, you cause injury to healthy cells.

How should a healthy immune system ideally handle inflammation?

The body makes both pro-inflammatory and anti-inflammatory compounds, both called prostaglandins, so the body can ramp up an inflammatory response if it needs to, and also dial it back in order to not burn down the house. If you have too high of an inflammatory response, you have an increased risk of heart disease, diabetes, cancer, and so on.

The goal is to have the right balance — have an immune system that has inflammatory compounds in its arsenal so it can defend against cancer and chronic diseases, but also to have anti-inflammatory compounds to suppress inflammation when it's not constructive.

And what's food's role here?

Food contributes to the body's production of prostaglandins. So an anti-inflammatory diet, as described on page 61, can help keep inflammation in check.

SUGAR

Should I worry about the high sugar content of fruit?
No. Although we understand why you're asking that, since there've been two major fads that renounced fruit in recent years. The first was the glycemic index; the second, fear-mongering about fructose.

What's the glycemic index?
The glycemic index is a measure of how much a fixed amount of sugar in various foods raises our blood sugar. High blood sugar, which can lead to diabetes, is a bad and dangerous thing.

And fruit has a high glycemic index?
There's a relatively high glycemic index for some fruits, and even for some vegetables. The conclusion was that these fruits drive up blood sugar, and the next conclusion was that fruit must increase the risk of diabetes.

That contradicts all your principal ideas about good, plant-based diets including fruits!
Not to mention that the fear over glycemic index also extends to vegetables like carrots.

Well, is the glycemic index a reliable way to judge the healthfulness of a food?
No. The measure has value, but judging a food based only on the glycemic index would be like judging a person's character based solely on, say, shoe size or bowling average. The glycemic index measures how much blood sugar goes up from the same amount of sugar in different foods. The problem with assigning that measure to fruit or carrots is that to get to the same amount of sugar from carrots and ice cream, you need a very tiny serving of ice cream versus a lot of carrots.

It doesn't account for the *concentration* of the sugar. This is the crucial distinction. Imagine, for instance, if we told you one person weighed 100 pounds and another weighed 200 pounds and then we asked you: "Who's overweight?"

Obviously the 200-pound person.
But what if we now told you that the 100-pound person was a five-year-old boy, and the 200-pound was a 6-foot-5-inch-tall man. That turns the tables: The man is very lean, while the child is seemingly obese, and thus "heavier" relative to height. Sugar is present in carrots, but it's pretty dilute. It's concentrated in ice cream. The glycemic index misses that distinction.

Is there a better way to understand the sugar content of foods?
Yes: the glycemic load. That *does* account for the concentration of sugar in a food, essentially adjusting for sugar concentration the way the body mass index (BMI) adjusts weight for height.

Does focusing on the glycemic index lead to any positive health effects, like weight loss or lowered blood sugar?
No. Studies have found an association between routine consumption of whole fruit and *lower* risk of obesity and diabetes. No one becomes obese or diabetic from eating carrots! We are generally making a mistake when we become obsessed with just one measure, any one measure, and think it's all that matters about a food.

You also mentioned fructose. Is it bad for you?
Our fixation on fructose is a variation of our obsession with the glycemic index. Many people started highlighting the potential harms of fructose. But fructose is fruit sugar. So it's in

apples, bananas, peaches, and every other fruit. And many vegetables as well.

But high-fructose corn syrup is bad, right?

While fructose is a natural sugar found in plants, high-fructose corn syrup is processed in factories by extracting and mixing fructose and glucose from corn. There are different versions, but on average the high-fructose corn syrup in the food supply is about 55 percent fructose, 45 percent glucose. Sucrose, or table sugar, is 50:50.

> High-fructose corn syrup doesn't occur in any natural food, so it is always a marker of a highly processed food.

But while there are differences in how they are metabolized, high-fructose corn syrup and table sugar are more alike than they are different in their effects on health.

Then why have we been led to believe that fructose is worse for us than sugar?

The usual: a bit of legitimate science, a lot of hype and distortion. Some particular harms have been associated with fructose relative to glucose — notably, the production and accumulation of lipids in the liver, leading to the condition known as fatty liver. This, combined with the quantity of high-fructose corn syrup in the food supply, led to a backlash against it. That, in turn, got distorted into the idea that *all* fructose was bad, and that all foods containing fructose were bad, and the next thing we knew, people were renouncing fruit!

What's the truth about fructose?

The truth about fructose is a lot like the truth about every other nutrient: It's the food that matters. Fructose on its own is generally found in fruits and vegetables, and those, of course, are almost invariably good for us; they defend *against* obesity, diabetes, and fatty liver.

High-fructose corn syrup doesn't occur in any natural food, so it is always a marker of a highly processed food. That food is apt to be bad for you for any number of reasons, the high sugar (fructose) content among them. The same is true of *any* added sugar, like sucrose, in a processed food: A load of added sugar is going to be bad, and may well be accompanied by other liabilities.

But whole fruit is good for us, and the fructose in it does not change that. We don't extract nutrients from food and live on those; we eat food and then the body processes all the nutrients in them. Eating fruit is not eating fructose; it's eating fruit.

So fructose is like other sugar?

The person most famously associated with chronicling the particular dangers of fructose, endocrinologist Dr. Robert Lustig, never recommended that people give up whole fruit. His particular concern was high exposure to fructose from added sugar, both high-fructose corn syrup *and* "regular" sugar.

Is there a good reason to use sugar substitutes?

Well, they're intended to take sugar out of your diet, take calories out of your diet, and not stimulate an insulin response, therefore — in theory — reducing the risk of weight gain, obesity, and diabetes.

Do they work? Do they do all those things?

It's not entirely clear. One of the biggest concerns about sugar substitutes is that they do nothing to file down a sweet tooth. If anything, they feed it and help grow it into a sweet "fang." Because although they may satisfy the craving for sweet, artificial sweeteners are intensely sweet: They range in sweetness intensity from 600 to 1,300 times as sweet as sugar. So even as they satisfy a craving, it might be that they're *cultivating* sweetness cravings.

So I'd be better off drinking "real" soda?

The best thing to do would be to not drink any soda, and, for that matter, to not add any sweetener — whether sugar or artificial sweetener — to your coffee or tea.

But my coffee tastes so much better when there's a little bit of sweetness.

For now it does, but as we mentioned on page 7, you can retrain your palate. Rather than focusing on how artificial sweeteners are lower in sugar and calories, you might focus on changing your taste buds' sensitivity for sweetness.

Right, taste bud rehab.

Yup. Artificial sweeteners don't help with that.

We think of artificial sweetener like a Band-Aid. Of course, if you're bleeding from a cut, a Band-Aid may help. So, we don't rule it out as valuable. But it's a temporary solution. For that cut, the long-term solution is for it to heal. For sugar intake, the long-term solution is to "heal" your diet so it's not loaded up with sugar and other sweeteners. That way, you won't be eating much sugar and you won't need to rely on artificial sweeteners to replace what you are already avoiding.

Are there other concerns about artificial sweeteners?

The direct concern is that these artificial sweeteners are exactly that: *artificial*. They are foreign chemicals that aren't native nutrient components of food. Studies in animals have shown that artificial sweeteners can disrupt the microbiome.

What does that mean?

Your microbiome is the collection of bacteria that live in your

GI tract (see page 189 for more about this). With artificial
sweeteners, you put a chemical in your food and the bacteria
say, "Hey, we don't recognize this." They get upset.

How do these disruptions affect my health?
That's the particularly vexing part. The disruptions of the microbiome that result from artificial sweeteners have been
shown in several studies to be directly related to the development of insulin resistance. Insulin resistance is a precursor to
diabetes, one of the things that avoiding sugar helps us prevent. So, it's ironic that chemical substitutes for sugar may
contribute to the very problem they are intended to help us
dodge.

To be fair, the literature is still murky. There are some studies that show that sugar substitute–sweetened beverages and
foods do help people reduce sugar and calories in the short
term, so there may be some advantage. But longer-term studies show the sugar and calories cut out from one place sneak
back in someplace else. And that the sweeteners themselves
can do harm in other ways. So, there's reason to be cautious,
and to minimize your reliance on sugar substitutes.

Okay, so back to sugar: Is it possible for me to train my palate to become someone who "just isn't a dessert person" or who "only likes dark chocolate"?
In fact, yes.

How?
If you eat a diet with minimal additions of sugar, you can sat-
isfy your cravings for sweetness with a lot less of the stuff.
Why? Because you become a lot more sensitive to it. The more
sugar and sweetness you eat, the more it takes to feel satisfied.
The less you eat routinely, the less it takes.

Almost like a drug?
Yeah, sugar has been compared to any other addiction: The more you feed your craving for it, the more it needs. It's called tolerance.

So it's possible to be addicted to sugar?
It depends on how one chooses to define "addiction" — those views vary — but for all intents and purposes, yes, because one of the characteristic traits of addiction is tolerance. The more you get, the more you need. You may satisfy your craving for sugar in the moment, but you cultivate the craving over time.

So what do you recommend?
Like we said, taste bud rehab. Learn about all the places sugar is hiding in your diet where you don't want it in the first place — like marinara sauce and salad dressing — and pick versions of those that don't have all that added sugar (or make them yourself). Bottled tomato sauce may have more added sugar (relative to calories) than bottled chocolate ice cream topping.

That freaks me out a little. Who would buy that sauce?
A lot of people, because they don't know the sugar is there — but they like the way it tastes. Ditto for salad dressing, breads, pretzels, chips, and of course breakfast cereal — where sugar is, at times, the main ingredient. It's safe to say that there's added sugar in almost every highly processed food. Manufacturers know that because we have that sweet fang, these things appeal to us. You might be eating a variety of processed foods to get the sugar you crave even if you're using an artificial sweetener in your coffee.

Either cook, or look at the ingredient list on products. If there's added sugar in there, pick something else.

But that still leaves the question: When I do consume sweet foods and drinks, should it be artificial sweetener or the real thing? Diet Coke or regular?

As long as sugar occupies a small place in your diet, we would recommend the real thing; a Coke a week isn't going to kill you. But think of our mantra, "Instead of what?" You're really better off with a peach and a glass of water.

Are the new sweeteners like stevia any better?

"Natural" sugar substitutes, like stevia or monk fruit extract, aren't as intensely sweet as older artificial sweeteners. And in comparison to Splenda (for example), stevia is better because it isn't a totally foreign chemical that disrupts the microbiome. (It's even been shown to have an insulin-stabilizing effect.) And since eating less sugar is beneficial, these natural sugar replacements are promising. But the research isn't conclusive.

Is there any tried-and-true way to lessen my addiction to sugar?

It's a quantity problem, really. The more you push sugar to the margins of your diet, the better. Focus on the usual: vegetables, nuts, seeds, whole grains, lentils, beans, and fruit. You don't need any of the sugar that's in processed foods. Then, if you have an occasional soda or ice cream as a treat, that's fine, because it no longer matters very much — you've already solved the bulk of the problem.

So how much sugar can I eat?

Generally, the recommendation is that at most 10 percent of calories can come from added sugar. But 5 percent, or about five teaspoons a day, would be better. If you're thinking in terms of 2,000 calories per day, that's a small soda, or a teaspoon of sugar in each of five cups of coffee, or some ice cream, or sweetened yogurt ... We're not saying "don't eat sugar." We're saying "don't eat a lot of sugar."

SALT

Let me specifically ask, does salt cause high blood pressure?
Yes, it does. In recent years, there's been an active debate in the medical literature about salt thresholds for health, and whether we have to worry about getting too *little* salt. But for most Americans, too much is the issue.

Generally, what type of person should worry about their salt intake?
People who have or are at risk for high blood pressure, kidney insufficiency, or congestive heart failure.

For everyone else, salt is okay?
The scientific, sensible questions are: Do we know what the ideal level of salt intake is? And do we know, in particular, what the ideal level of salt intake is for different segments of the population: healthy people and people at risk for various conditions, like high blood pressure? Those are the questions we should be asking.

Well, do we know the ideal amount of sodium intake? And is "sodium" the same thing as "salt"?
"Salt" is a chemical term, referring to the combination of a positively charged ion with a negatively charged ion. Table salt is a combination of sodium and chloride, but it's so common that we use the words "salt" and "sodium" interchangeably — because almost all the salt we use (in eating, at least) is sodium chloride. Half of that salt is sodium, the other half is chloride.

Anyway, the conclusion from a flurry of studies over recent

80 percent of the salt in the typical American's diet comes not from a saltshaker, but from salt that's processed into food that you buy in a bag, box, bottle, jar, or can.

years is that we're not entirely sure what the ideal intake is. Intake of 1,500 milligrams a day may be too little, and 1,200 milligrams a day is almost certainly too little; 2,000 milligrams (2 grams, which is a substantial pinch) a day may be just right, and more than that is probably too much. But we're not entirely sure.

If we have no idea how much is a healthy amount, then how much sodium should I be eating?

In a headline-driven society, when we admit that we don't know what the optimal level of sodium is, some people conclude that we know nothing at all. Therefore, why restrict sodium? But all the reasons not to restrict sodium are highly theoretical, implausible, and detached from reality. Too much salt raises blood pressure; that we know.

What's the reality?

The current recommendation for salt intake in the US is up to about 2,400 milligrams a day for adults. But the average adult gets 50 percent more than that — over 3,500 milligrams. That's almost certainly too much, and lots of people in America consume over 4,000 milligrams; it's not unusual to be eating 5,000 or 6,000 milligrams.

This all sounds alarming, but what exactly do these numbers mean?

We tend to use 2,000 calories a day as a prototype for adult dietary intake. So 2,400 milligrams of sodium, which is the recommendation, would be, on average, 1.2 milligrams of sodium per calorie. If you check a label and the milligrams of sodium are way higher than the number of calories — that's too much sodium.

Can I eat too little sodium?

The recommended minimal salt intake in the Dietary Reference Intakes report from the National Academy of Medicine — 500 milligrams a day for an adult — may be too low for some people. (Remember, that's the low threshold.) But this idea of whether you are getting enough salt has been corrupted into a misleading, unhelpful discussion. The chances are you're getting more than enough.

How do I know?

Look at the nutrition facts panels on some of the things in your pantry; you may find that many have more milligrams of salt per serving than calories. Eat enough of that kind of stuff (and of the kind of food served outside of your house), and you're almost certain to get too much sodium.

In other words, if you eat highly processed foods, it's a challenge to hit the target of 2,400 milligrams a day and 2,000 calories — because most of these foods have more milligrams of sodium per serving than calories.

Okay, I'll stay away from salty snacks like potato chips.

It's not just salty foods: Like sugar, salt is put into almost everything, and liberally. But, oddly, the best foods in the salty snack aisle can be relatively low in sodium, because they have very few ingredients. For example, if you buy corn chips made from only corn, canola oil, and salt, every little bit of salt put in that very simple formula of three ingredients is on display; you can taste it. (We're not pushing corn chips, understand; we're pushing awareness.)

Salt hides in unexpected places. Foods like breakfast cereals almost always have a high concentration of sodium relative to calories. If you're thinking, "I'm having cereal for breakfast; that's not a salty food, so it's fine if I have deli meats for lunch ... " you'll probably be way above your recommended sodium average by the time you finish breakfast.

In general, packaged and ultraprocessed foods are where your sodium is coming from: 80 percent of the salt in the typical American's diet comes not from a saltshaker, but from salt that's processed into food that you buy in a bag, box, bottle, jar, or can. Restaurant food is also notoriously salty.

Why is sodium ubiquitous in ultraprocessed food?

For good reasons. One, salt makes food tasty. Two, it stimulates appetite. And three, it preserves food, increasing the shelf life.

Still: We need sodium, right?

Yes. It's a key constituent of blood. It plays many critical roles in regulating the function of the cell membrane. It's critical to the nervous system.

Is there an easy way to make sure I'm not getting too much salt?

The DASH diet (see page 58), which was designed to reduce blood pressure, will do that if you stick to it. Or: Avoid ultraprocessed food and don't eat out too much. That'll probably do it also.

ANTIOXIDANTS

Do antioxidants deserve all the attention they get?

Yes! Antioxidants play a major role in minimizing damage to cells. We are exposed to rogue cells, dying cells, and potential invasions from dangerous pathogens in the environment all the time. Antioxidants protect healthy cells from being injured.

How do I know if I'm getting enough antioxidants?

You get a rich infusion of antioxidants if you have a good dietary pattern. But the standard American diet is low in antioxidants. On top of that, ultraprocessed foods can cause the very issues that antioxidants fight against.

So eating a lot of ultraprocessed foods and not enough antioxidants is a dangerous combo?

It means that you're doing two things that are potentially injurious to healthy cells: 1) You're eating inflammatory foods and increasing the amount of chemical attacks; and 2) you're reducing the availability of defense mechanisms that shield healthy cells from these chemical weapons. So there's much more collateral damage being done. And, not coincidentally, eating a lot of ultraprocessed foods probably means you're not getting enough antioxidants. And getting enough antioxidants probably means you're not eating a lot of ultraprocessed foods.

Are there different types of antioxidants?

Antioxidants live in two major families: carotenoids and bioflavonoids. Both are large groups of structurally related compounds that help plants cope with the radiation in sunlight. When animals eat plants, these compound help squelch the damage that "oxidants" — or products of oxygen — can do to cells. They help animals cope with sunlight, too, and other stressors that can injure cells. Think of oxidation like cellular rust, and think of antioxidants as rust-proofing.

What foods are rich in antioxidants?

A whole variety of fruits, vegetables, whole grains, beans, lentils, nuts, and seeds. Carotenoids tend to be concentrated in

brightly colored fruits and vegetables, such as carrots, toma-
toes, and oranges. Bioflavonoids are particularly concentrated
in tea, wine, apples, legumes, and berries.

VITAMINS AND SUPPLEMENTS

**How about supplements? Everyone I know takes vitamin D,
for example.**
Vitamin D, unlike most nutrients, is actually a hormone with
the complex name of 1-25 dihydroxycholecalciferol. The compound results when our body manufactures vitamin D in the
skin with exposure to sunlight, and then
activation steps occur in the kidneys and
lungs. From there, among other things,
vitamin D determines whether we absorb calcium when we ingest it. That's
why dairy is fortified with vitamin D:
The calcium in dairy is almost useless without it. We've known
that for a long time, but other effects of vitamin D, like the
modulation of immune system activity, have been discovered
as well.

> Supplements should
> be supplements to
> a good diet, not
> substitutes for one.

**Does vitamin D have the same effect whether you absorb it
from sunlight or drink it in fortified milk?**
Pretty much. All that really matters is that you are getting
enough of the active form.

Why do we fortify food with vitamin D?
Because we don't get as much sun exposure as we used to.
During the Industrial Revolution, when children began working in factories from dawn to dusk, they stopped producing vitamin D, stopped absorbing calcium, and began to suffer from
rickets — distorted bones. When milk was fortified with vita-

min D, the rickets epidemic went away — it's considered one of the triumphs of modern public health and modern nutrition. So to this day, dairy and many other foods are routinely fortified with vitamin D. Vitamin D supplementation is another approach, and often recommended, because while rickets is now very rare, suboptimal levels of vitamin D in blood are fairly common.

I thought I'm not supposed to eat processed foods, and I'm supposed to cut back on dairy?
There's vitamin D in many animal foods, including dairy and some fish and meat. And although adding a nutrient to a nutritious food like milk or yogurt is a form of processing, it's not a form we have a problem with. The keys are: The food must be worth eating to begin with, and the nutrient must be of probable, genuine benefit.

How do I get my vitamin D if I'm vegan, or lactose intolerant, or simply want to eat a diet that minimizes animal foods?
If you don't eat meat and dairy, you're not getting sun exposure, and you're not eating fortified processed foods, you should take a vitamin D supplement. Which is certainly as useful, and perhaps more so, than drinking milk to get vitamin D.

And if I spend a lot of time outside and eat meat, dairy, or fortified foods, I'm all set?
Any combination and you're all set. If you are out in the sun routinely (without sunscreen), you don't need vitamin D in your diet at all. (Generally, exposing arms, legs, and face to direct sunlight for 20 minutes over the course of a day is ample. That is obviously easier in some climates than others, and at certain times of year.) If you get vitamin D from a variety of

dietary sources, with or without sun exposure, you are probably going to get enough. If in doubt, talk to your doctor about
checking your level. There's no harm in taking a vitamin D
supplement.

Taking a higher-dose vitamin D supplement in winter than
in summer can make sense, since it is likely you are out in the
sun more in summer and consequently making more vitamin
D naturally.

Do I need to be taking other supplements?

We often use the word "supplement" without bothering to
ask, "Supplemental to what?" Let's start with this: Supplements should be supplements to a good diet, not substitutes
for one. They're not an excuse to eat less nutrient-dense food.

Okay, so you're anti-supplement. Got it.

No. We talked about vitamin D: Most people should supplement that, at least in winter. In general, if you aren't getting
important nutrients from your diet, you should supplement.

How do I know which nutrient supplement I should be taking?

Before you start thinking about what nutrient supplements
you should take based on something you read online, think
about your dietary pattern. Do you know the overall quality
of your diet?

If you're very active and you have an optimal diet of
high-quality, whole foods, the likelihood of you needing supplements is small, because you're getting a wide array of vitamins, minerals, fiber, antioxidants, and so on in your diet.

What type of person would benefit from supplements, then?

On the other hand, if you eat very few calories — which would be true of homebound elderly people, as one example — even if your diet is high-quality, you may be eating so little that you're not hitting optimal threshold values for many nutrients. Some people may have trouble absorbing a specific vitamin, such as B12 or folate; it's something your doctor can test for with blood work during your physical.

Since I don't exactly know what type of supplements I should add to my diet, I always take a multivitamin. That can't hurt, right?
You could definitely argue that a multivitamin mineral mix is an insurance policy for making sure you aren't deficient. However, it's not clear that multivitamins or any other supplements are directly linked to health consequences. There are some studies suggesting a modest net benefit of taking supplements; but there have also been some studies that raise the possibility of harm.

So there's no proven benefit associated with taking some vitamins?
If only it were that simple. The potential for harm is fairly remote, but so is the potential for meaningful benefit for most people. (It's worth pointing out that supplements cost money, too.)

Why can't you say conclusively?
Here's the thing: Supplements are missing many of the nutrients that are found in whole foods. There are thousands and thousands of nutrient compounds in plants, and there are only a few dozen nutrients that are isolated and put into even the most complete multivitamin mineral formula. So taking a supplement just isn't the same thing as eating highly nutritious food, like broccoli or an apple.

What about whole-food supplements? Those must be a little closer to eating whole fruits or vegetables.

Allegedly, whole-food supplements are the equivalent of eating multiple servings of fruits and vegetables. The argument here is that if you are not getting the recommended daily intake of fruits and vegetables, you could take an encapsulated concentrate of essentially dehydrated fruits and vegetables, in which all the native nutrients in those plants are preserved. The fiber and pulp is taken out, so it fits in a capsule. It's an intriguing theory, premised on the idea that nutrients in foods act in concert with one another; and that if we can preserve that native arrangement of nutrients, we may be more likely to see a benefit than we would if we isolate the nutrients into a multivitamin.

> Remember that if you're eating a lot of actual fruits and vegetables daily, that makes up the bulk of your diet, which means you are eating less other, unhealthy stuff.

But though the theory is valid, the research doesn't yet answer these questions definitively: There's no evidence that these supplements produce any of the same outcomes as eating fruits and vegetables. Remember that if you're eating a lot of actual fruits and vegetables daily, that makes up the bulk of your diet, which means you are eating less other, unhealthy stuff. Supplements don't lead to that same pattern.

If you're taking these whole-food concentrates to get those same nutrients, it's presumably because you are *not* eating a lot of fruits and vegetables daily, and instead are eating ultra-processed foods and/or animal products. The effect of all that other stuff is still in play even if you are taking a supplement.

In short, you can't just count on the "nutrients" in fruits and vegetables to do you good. You count on the *actual* fruits and vegetables to do you good. Worry about the food, and the nutrients will take care of themselves. Worry about the nutrients, and you may get it all wrong.

Okay, let's get to the point. Are there supplements most people should take?

We could tailor supplements for each individual's dietary pattern; the primary advice here is about a healthy diet. The shortlist for supplements that most everyone should take comprises vitamin D and omega-3, which can come from fish oil or vegan algae.

What about probiotics, which seem to be in vogue?

Those are a good option to consider, especially because a lot of the exposures of modern living — from chronic stress to sleep deprivation to food chemicals — have the potential to disrupt the microbiome (more about that on the following pages). Probiotics help restore healthy colonies of bacteria in the gut.

What's the bottom line?

Supplements are a multibillion-dollar industry, and a lot of that is marketing. But the argument that supplements are useless and everybody should stop wasting their money is also overblown: There are gaps left by the typical American diet, and those gaps can be plugged with the judicious use of supplements. But if your diet is good, the supplements you take should be minimal.

THE MICROBIOME

Earlier, you mentioned the microbiome. What is that?

"Biome" means "living entity"; the term "microbiome" refers to our personal colony of microorganisms, mostly bacteria, configured in our body. Everyone has one, and it's home to a lot of bacteria (in fact, the bacteria in the human body outnumber human cells by an order of magnitude).

That's quite amazing. What does it do?

The microbiome is crucial to our digestion and the integrity of the intestinal lining, and to determining how and when and where things are absorbed into the bloodstream. It participates actively in our metabolism and it plays a role in our immune defenses. In the gastrointestinal tract, the bacteria in the microbiome digest things that we couldn't digest otherwise.

The influence of the microbiome reverberates throughout our health. Whether it is healthy or disrupted has implications for our defense against infectious and chronic diseases, and our susceptibility to insulin resistance, diabetes, coronary disease, and more.

What can disrupt the microbiome?
The microbiome is disrupted by the antibiotics and pesticides in our food, as well as by artificial flavorings, artificial colorings, artificial sweeteners, and more. It could also be disrupted simply by the high concentrations of things like sugar so prevalent in ultraprocessed foods. Diets high in meat result in markedly different microbiomes than plant-predominant diets, too. In a way, you could say that almost everything that makes the modern food supply what it is today is disruptive to the "ancient normal" for the human microbiome.

Is every microbiome different?
Yes, and each one changes. When the bacteria that make up the microbiome shift this way or that, it changes the relative concentrations of a wide variety of metabolites — compounds that result from the activity of the bacteria and wind up in the bloodstream. This conversation takes us in the direction of specific compounds, like TMAO, which has been shown to interfere with endothelial function — the ability of your blood vessels to constrict and dilate (blood vessels that can't dilate when they should can cause a heart attack or stroke).

What foods contain TMAO?

TMAO is not itself contained in foods, but rather is produced by bacteria when they digest amino acids like choline and carnitine. Meat, a concentrated source of carnitine and choline, is the best-known culprit. As we mentioned above, people who routinely eat meat have a different microbiome than people who don't, and their microbiomes produce more TMAO.

So if I eat meat, I will most certainly have TMAO in my bloodstream?

Yes. But it also depends on what bacteria your microbiome already has.

What if I'm vegetarian?

A vegetarian who occasionally eats meat will produce less TMAO than a routine meat-eater, thanks to the long-term effect of their habitual plant-based diet. How much TMAO you make seems to depend both on how much of the sources you are eating at a given time and how you have shaped your microbiome over time. TMAO is just one example; there are many others.

Sounds like my microbiome is one more part of my body I have to worry about! How do I treat it right?

It's true that the microbiome is an important aspect of our health, but the idea that we need to eat to feed our microbiome takes things a bit too far. If you're feeding yourself well, you're feeding your microbiome well.

Just as you don't eat to feed your kidneys or your liver — both essential to your health — you don't need to eat for your microbiome. It's a vital part of you, and your health, one that deserves to be understood and respected. But we don't have to give it special treatment.

So there's no different formula to eat for a healthy microbiome? The same knowledge of a healthy, plant-based, balanced diet still applies?

We already know that people who eat a balanced, whole-food, real-food diet, exercise routinely, don't smoke, get enough sleep, manage their stress, and have good social interactions have way lower rates of chronic disease and enjoy a vital life. When we suddenly learn that the microbiome is important to health, we don't have to revisit all of that.

That said, there's some tinkering you might do, because the microbiome also has a role in your resting energy expenditure (REE), the amount of calories your body uses when it's at rest. Variations in REE are among the reasons two people can eat the same and exercise the same, but one gets fat and the other one stays thin. While genetics is a main contributor here, it's possible that shifts in the microbiome can influence the number of calories required to just exist. When somebody finds it stunningly hard to lose weight or agonizingly easy to gain weight, their microbiome may be part of the equation. That does suggest that at times direct attention to the microbiome with diet or supplements, such as prebiotics and probiotics, could be important. But that's a very individual thing.

Anything else?

Well, since it's also responsible for a lot of digestion processes, people who have digestion problems may want to look at shifting their microbiome. Again, this isn't everyone; it's part of addressing a specific problem.

And if my microbiome isn't cooperating?

It can be altered with probiotics (bacteria you ingest), prebiotics (nutrients to help support bacteria in the gut), or antibiotics (to kill bacteria if the wrong ones are prevailing). An area of active research now is fecal transplant, which means importing a healthy microbiome to replace an ailing one. You can fill in the rest of what it means with your imagination.

If you have gastrointestinal issues or indigestion, if you are intolerant of some foods, if constipation is an issue, or if you have vague symptoms following meals — like fatigue and brain fog — it may mean that you are not optimally digesting and metabolizing food. So taking a probiotic and seeing whether or not it changes how you feel is worth trying.

Can I take probiotics as a preventative way to keep my microbiome happy?

Yes, that's another argument for probiotics. They can be used as a precautionary measure to replenish the microbiome. The truth is, it's nearly impossible to eat food in the modern world and not get some antibiotics (which disrupt the microbiome) in there, because they are so widely distributed in the environment.

QUESTIONING THE ANSWERS

(On Science and Sense, or,
How We Know What We Know)

On Research

A lot of the headlines I see about nutrition science make me feel like I'm doing everything wrong. What's actually happening in nutrition science right now?

There's an old parable about six blind men who encounter an elephant. Each touches only one part — side, tusk, trunk, knee, ear, and tail — and so each imagines the elephant to be a different shape (a wall, a spear, and so on). Of course, without seeing the complete picture, none of them is right.

In an age of avocado toast and oat-milk lattes, it's hard to see what that has to do with nutrition science.

It's relevant because we're prone to a similar tendency in nutrition science, which is intrinsically reductionistic: that is, blindly analyzing pieces of the whole in hopes of putting together an accurate big picture. That's what methods like randomized controlled trials do: People are randomly assigned blindly to different groups in order to reduce bias. That's justifiable because one goal of science is to peer into the spaces we can't see with our unaided senses. So we've developed microscopes and telescopes, as well as methods to reveal things that the unaided senses can't perceive.

But it isn't our only method of knowing things. While science is the most powerful means we've developed to answer hard questions, the source of the good questions in the first place is sense.

So science starts with sense?

Exactly. And being sensible means remembering that many

questions are answerable without science. We might need science to answer the question, "Why does gasoline cause fire
to burn faster and hotter?" but we don't need it to answer the
question, "Should I try to put out this fire by pouring gasoline
on it?"

Sense tells us stuff that we know perfectly well without any
particular kind of scientific method or instrument. We know
that water is better for putting out campfires than gasoline.
We know that getting shot through the chest is bad for people and emergency surgery for fixing the bullet hole is generally their best hope — we don't need a randomized controlled
trial to know that. We know that if we toss an apple up into
the air, it will fall back down. And we know that eating that
apple is generally a good idea. Again, no randomized trials required.

What does that have to do with modern nutrition science?

It's very easy to ask a bad question and design a trial to answer it. And there are no good answers to bad questions. Just
because you're running a randomized controlled trial doesn't
mean you're going to find out something useful.

Give me a real example of a bad research question.

Even the question "Which is better, a low-fat diet or a low-
carb diet?" is a bad question because sense tells you that there
are good and bad ways to put together a low-fat diet. You can
have a low-fat diet that just happens to be low in fat because
it's mostly made up of very wholesome, whole plant foods. Or
you can have a low-fat diet that is made up of sugar-sweetened
beverages and cotton candy. They're both low-fat, and they're
night and day.

A low-carb diet can be a diet that's rich in highly nutritious foods. A low-carb diet can also be unhealthy and full of ultraprocessed fat and added sugars, or a diet that is all fatty meats and completely devoid of fruits, vegetables, beans, lentils, whole grains, and all the most nutritious foods. When you ask a silly question like "Is a low-carb diet better?" you can put together the comparison in any way you want to guarantee the outcome you are looking for. And that kind of thing happens all the time in randomized controlled trials.

> When you go looking for the active ingredient either to account for the actual dangers of a food or the potential beneficial effect of a food, you miss the forest for the trees.

But there seem to be contradictory conclusions for literally every food. It jumps from "beans are toxic" to "wait, just kidding, beans are awesome and you should eat them every day."

Some conclusions are only contradictory when you study one isolated compound at a time. It would be like saying: "Okay, we want to know the net effect of breathing. We understand that the most important component of life to people is oxygen and that too much carbon dioxide is bad, so we're going to concentrate and purify oxygen, get rid of those nasty other gases, and study the effects of that on people." Well, the effect of breathing pure oxygen is that you're dead within three days.

Now oxygen has become a toxin. Not just a toxin, but one that kills you pretty quickly. Since air contains oxygen, then, we'd better hold our breath. That's the sort of logical conclusion you are obligated to reach if you follow this kind of thinking. When you go looking for the active ingredient either to account for the actual dangers of a food or the potential beneficial effect of a food, you miss the forest for the trees. We've said this before: Eat food, and the nutrients will take care of themselves.

But aren't the component parts of food—vitamins, minerals, the whole shebang—what defines whether a food is healthy?
Here's another example of why simplifying anything into its component parts doesn't give you the full picture. If you can see perfectly well that a levee is useful in holding back the rising waters of a river spilling over its banks, you could apply reductionistic science to the question: "Gee, what sandbag in that levee is the active ingredient?" But there is no active ingredient other than the levee.

In nutrition science, the reductionistic question goes like this: "What is the active ingredient in broccoli that is responsible for its health effects? Is it antioxidants? Vitamins K and C? Folic acid? Magnesium? Fiber?"

That dietary patterns are profoundly associated with health outcomes, and that broccoli is a part of many good dietary patterns, isn't satisfying for people who use reductionistic thought. However, it seems likely that the active ingredient in broccoli is broccoli. Broccoli is the levee.

Well, that's just an unsatisfying answer.
It's unsatisfying if you want to be able to break things down and pick them apart, and Western thought has led most of us to think that way. But that kind of thinking is not always useful, and it's often wrong. A lot of nutritional research goes into looking for that active ingredient, and that can be useful. But sometimes when you focus on isolated components of foods, you invite egregiously misguided interpretations. The "broccoli" answer is deeply satisfying if you want dietary advice that's uncomplicated and easy to follow and build a healthy life around. That should make you overjoyed!

THE FOREST VERSUS THE TREES

So what's the sensible way to approach nutritional science?
As we said, it's simple: If you define diets by foods, it's much
harder to get it wrong. At some point, it doesn't require the results of a particular study to say that good diets are made up of
wholesome foods. It's sensible. The notion that you can define
the quality of a diet on the basis of a particular nutrient level
is misguided and antiquated, and it needs to be abandoned.
It's really about the food as a whole, not food's component
parts.

The whole is what the whole does.
Exactly! And what you hear about the part doesn't change
what the whole does. So you could say breathing the air of this
planet is our best option even though it contains oxygen and
even though oxygen in high concentrations is toxic. Maybe it
would be better if there were some alternative, but there isn't.

**That sounds like a problem, that the purpose of science has
been corrupted . . .**
In the case of science and diet, scientists have actually talked
themselves into unknowing what we formerly knew, and knew
forever — so for example, we know that fruits and vegetables
are good for you. It's a fact. That this can be questioned is not
really a problem of science, but of the interaction of science
and sales, media, pop culture, the web, and the fact that (le-
gally) anyone can say anything and if it's outrageous enough,
many people will see it.

How we feed ourselves is as much about sense as science,
and it's only recently — with the advent of non-food foods (or
unidentifiable food-like objects) that it's become confusing.
Just because you put something in your mouth and chew and
swallow it doesn't make it actual food.

Why would we decide to unknow things in science rather than focus on striving for new knowledge?

We don't need to create randomized controlled trials to prove in a less controversial way that fruits and vegetables are good for people. (Or even, really, whether a strawberry is "better" than a blueberry.) The more important thing at this point is to try to figure out how to get fruits and vegetables to be a more routine part of everyone's diet; how to make them more available, more ubiquitous, more affordable; and how to get everyone to eat them routinely.

What do manipulative studies look like?

If you're a champion of low-carb diets, you devise an experiment that pits a good low-carb diet against a crappy low-fat diet. If you're a champion of low-fat diets, you do the opposite. If you're wondering, "How is it that we can get all these different trials that seem to be saying completely different things?" it's because sense was never in the driver's seat. We asked bad questions and used science to generate answers. And there are no good answers to bad questions.

How do I not get duped by these kinds of studies?

The important thing is to not get carried away by any one given study, *especially* if it's an outlier. It's unlikely at this stage that we're going to see anything new and really novel, out of the blue, in the world of diet and nutrition. Drinking soda is not suddenly going to become beneficial, or eating broccoli bad for you. Yet we could design studies that would "show" these things.

Are there other, more reliable ways to learn about nutrition?

Absolutely. The truly reliable way is to look at the big picture —to see the elephant—not just a piece; to look across the broadest expanse of evidence, including but not limited to intervention trials, what we know about our own native past, and what we can observe about the fate of whole populations over generations. Different kinds of evidence are like different pieces of the same, great puzzle — you need lots of different pieces to complete the picture.

It's critical to consider knowledge that existed before we invented the randomized controlled trial. There were clear patterns in populations: There were populations with traditional ways of eating associated with low levels of chronic disease, and populations with ways of eating that were associated with high levels of chronic disease.

Take beans, which are a staple in all five of the healthiest known populations in the world (sometimes called Blue Zones), places where's it's commonplace for people to live to be a hundred. From these places, we know that if you want to improve your diet, you should routinely eat beans. *We don't need a study for that.*

Can you really study how a population's unique diet and culture affects overall health?
We have natural experiments, like cultural transitions. In India and China, for example, traditional ways of eating were plant-based and consisted of simple, real foods and a reliance on grains and legumes. Both populations had limited meat intake. They drank tea, not soda. And guess what? There were generally very low rates of chronic disease. People were prone to nutrient deficiencies if they didn't have dietary adequacy, but they weren't getting chronic disease. In the last generation or two, globalization led to an influx of McDonald's, Coca-Cola, Dunkin' Donuts . . .

And then . . . ?
Consumerist culture mixed with an intentional, almost dictated disconnect from traditional food systems. The concentration

of people in cities, the end of people working and living off the land, led to the "opportunity" — the necessity, really — for urban Indians and Chinese to begin to eat ultraprocessed food, fast food. All of a sudden, almost in the blink of an eye, rates of diabetes and obesity and heart disease exploded. We saw the same thing in the Middle East, where oil led to "modernization" and an influx of Western comforts. We saw both reduced physical activity (which, remember, even though we're not talking about it much, remains important) and increases in ultraprocessed food. Again, almost immediately, we saw higher rates of type 2 diabetes.

So what does that tell you?
Generally, recognize that your diet is what you eat and drink over the course of a day, and that when you change that, when you add something or take it away, that change reverberates through your dietary pattern.

How significant is each change?
On page 167 we use the example of eating eggs — are they good for you? And it's tricky: It's not a yes-or-no question. If you weren't eating eggs before and you start eating eggs, one of two things has to happen. You're either adding eggs on top of everything you ate before, in which case your calorie intake has gone up; you probably don't need that. And now you are looking at two variables — eggs and total calories. This is problematic because the point of much research is to try to isolate one factor you can study.

Or you're adding eggs instead of something else. Are the eggs an upgrade or a downgrade? Did you replace a plate of lentils, or a double cheeseburger? If you ate eggs instead of cheeseburgers — awesome. If you ate eggs instead of lentils, you've added saturated fat and removed fiber.

The fact that you are eating eggs can't be isolated; you've got multiple factors in motion, so whatever you see happening is attributable to one or the other or both.

Jeez, there are so many ripple effects of adding one measly egg to your diet. It's overwhelming.
You can make minor changes in a generally good diet without driving yourself crazy. Take a breath. Eat the egg.

Generally, you've got a couple of options. You can make one-for-one substitutions: Stop eating bagels and eat eggs instead. Or you can shrink the portions of everything else to make room for the eggs. Of course, now you've shifted the percentage of calories derived from different foods. As we said, don't make yourself crazy; just try to make upgrades in your diet, not downgrades. Don't substitute double cheeseburgers for eggs.

This must make it hard to create a sensible research question.
Yes and no. We tend to conflate the reliable evidence we can generate about what happens to people with different dietary patterns with what we can learn about shifting single foods, single ingredients, single nutrients. Which is really much harder to do with studies of diet. Any time you change any aspect of diet, you inevitably end up changing something else.

But you're saying it's difficult to account for one thing, like adding an egg a day?
Researchers are bogged down in the "what." What is the thing that is changing health outcomes here? Is it eating more eggs? Or eating fewer donuts? A study that's looking at the same isolated aspect of diet could lead to different conclusions. You just took away something that in large doses is akin to poison

and substituted real food. Which has the greater effect? Maybe we'll never know. But we know it's better to eat fewer things that diminish health, and more real food, which enhances it.

It's much easier to go back to the original question at the heart of this book: "What constitutes a good diet?" That's a way more important question than "Can I eat an egg?" No individual food is going to kill or save you, unless you make it a major part of your diet.

So the net effect of eating eggs isn't really about the eggs at all.

Well, it's not *only* about the eggs. Because, as you see, there could be two studies entitled "The Effect of Adding Eggs to a Diet" with two totally different conclusions. One could be that adding eggs to a diet offers a net benefit in overall satiety, weight control, and cardiovascular status. But another study with the same title could reach the exact opposite conclusion — that there is a net decline in overall cardiometabolic health, a degradation of diet quality.

> Almost never does a single study change everything we thought was true.

And these studies (which are much more complicated than they appear in a title) immediately pop up in the headlines, I assume?

The whole truth — that is, in this case, that it matters if eggs are replacing donuts or oatmeal — gets lost in a twenty-minute cycle of hyperbolic headlines. You're going to see "Eggs at Breakfast Show Benefit" or "Eggs at Breakfast Increase Disease." That's why you're confused, and so is almost everyone else. You gotta ignore the headlines.

Isn't science all about making big discoveries?

No, it's all about filling in gaps — because *science is incremental*.

This is a critical point. *Almost never does a single study change everything we thought was true.* You're talking about Einstein when you're talking about that. Almost inevitably, even the largest and best studies only make sense when they interpret the context of what we knew before. If you sort through all this, you realize that we're being told erroneous things about how we use science. Our entire culture is complicit in this. Scientists themselves wanting to be heard, the media wanting to capture our attention, and us thinking that every study is a fundamental new truth rather than an incremental contribution to the weight of evidence. "How does this add to what we already know?" is a good question. "Does this study change everything we knew before?" is idiotic.

The thing we tend to forget is that there are so many of these studies from two years ago, three years ago, five years ago, and — what now seems like ancient history — ten years ago, that are every bit as robust as studies that make headlines today.

What red flags can I look for when it comes to nutrition science in the news?

It's a red flag when an article claims to have discovered something that is a fundamental change in what we knew about nutrition. Like that beans are all of a sudden toxic (which we debunk on page 96). That sounds so dumb, it's hard to believe the statement could get published — beans are, in fact, the world's most important (and healthiest) protein source — and yet studies have "found" that very idea.

What should we be looking for instead?

We should be looking for studies with incremental contributions. That answer both questions: "Is there a dietary pattern where an addition of eggs confers a benefit?" and "Are there other dietary patterns where the addition of eggs confers harm?" Those are legitimate questions. A study should build upon the basic set of already clear truths we know, truths that are clear on both the basis of science and sense.

RESEARCH METHODS:
ONE SIZE DOES NOT FIT ALL

Relying on the randomized controlled trial seems the opposite of what you're advocating for: that we should be embracing what we already know about nutrition.

It can be both. The idea is to be careful about any new randomized controlled trial, to make sure it tells us something valid and useful. When it does, we want to add that to the foundation of prior knowledge — not explode the foundation and start over. The intellectual contrarians fabricate heated debates about diet in such a way that assumes we never know anything unless we know it from a randomized controlled trial. That's patent nonsense. Again, we don't know that water is better than gasoline for putting out a fire from a randomized controlled trial. But anybody with sense would say, "Of course gasoline makes campfires blow up! Of course Cheetos are bad! Of course kale is healthy!" And while we are at it, we have no randomized controlled trials to prove the harms of smoking, either, a fact the tobacco industry used to delay reaction to the obvious, calamitous toll of tobacco use.

Can you break down how a randomized controlled trial works?

Sure. They are experiments that are supposed to be unbiased. Of course, this doesn't always happen; but generally speaking, they are randomized, meaning you can't cherry-pick participants for one group versus another. You don't know who is assigned to each treatment.

Do the participants know what group they are assigned to?
Nope, not when studies are "blinded." You are helping to control for placebo effects and related phenomenon. If people know they are being assigned to an active treatment, they assume they will get better. If they are assigned to a control group, they might not expect to get better.

The question would then become, "Is it the treatment that is making the difference or the expectation of the effect?" So randomized controlled trials are so popular because blinding is important. (Of course, not all randomized controlled trials can be blinded: You can't blind people to whether they are eating eggs or donuts.)

What are some of the limitations of randomized controlled trials?
For example, let's say you wanted to study the association between dietary patterns and cancer. What if you compared the effects of bacon, or kale, on two groups, but didn't consider the factor of smoking, and you inadvertently had one group whose participants smoked a whole lot and one that didn't smoke at all? If you are only looking at bacon or kale and cancer outcomes, you may conclude that diet in the first group — even if it's kale! — leads to cancer. But, in fact, it would be the smoking. Smoking is what we would call the confounding variable. Confounding variables are everywhere. Randomized controlled trials are very hard to perform with reliably intelligent results.

How do you prevent confounding variables, like smoking, from having an overall effect on the study?
Large groups and randomization, among other methods. The hope is that with a large enough group, you get an equal amount of smokers and nonsmokers. Variables like that are

very obvious. You can also limit eligibility based on important measures — only nonsmokers allowed, for instance — or acknowledge such variables and explicitly build them into your analysis. But sometimes there are genetic mutations or other factors that we don't even think to measure, and that's where randomization makes its unique contribution.

Does a large group guarantee reliable results?
When you randomize — try to distribute random people in different groups — you can't guarantee an exact balance. In large randomized trials, there is a very high probability that all the factors we know are important and all the factors we don't know are important will be pretty balanced between both groups.

What is considered a large enough trial?
You can capture a lot of what matters with trials of thirty or so much of the time. But the larger the randomized trial, the larger the likelihood of achieving balance. This also obviously depends on the nuance of the particular research question, which is where our best friend — sense — enters the picture.

There is a purely statistical element here, too: You need a large sample when there's a lot of variation in responses, and the average effect of treatment is very small. The converse — a large treatment effect, and a great deal of homogeneity in that effect — allows for a clear view of what's true with a small sample.

What can't randomized controlled trials tell us?
Long-term effects, for one. Consider, for instance, whether or not you would be willing to be randomized for the next thirty years of your life to a low-fat diet or a low-carb diet, to a vegan diet or a ketogenic diet, or any combination.

Or you may be assigned for the next decade of your life to eating no fish at all, to eating a wide variety of fish, to only eating swordfish, that sort of thing. It's not like, "Here, live your life and either take this actual pill or this placebo pill and don't worry about it." It's "We want to change how and what fish you eat every day for the next decade." You up for that?

That's a major life commitment; I don't know if fish and I are ready for that step in our relationship.
You're not alone. It's not just that these studies would need to be very long, very large, and stunningly expensive, it's that very few normal people would be willing to sign up for them in the first place. If you did get people to sign up for a study that arduous — who are these people? What normal, reasonable person is willing to do that? And if they're not normal, if there's something just a little off about these people, is what we learn about them referable to the rest of us who are obviously so robustly normal?

In theory, you could run the trial in the gulag or someplace where you tell people what to do and they have no choice. Prisoners.

Is that ethical?
No. And believe me, those issues are a huge historical problem in research. It's been done in the past, and it's one of the reasons we have such strict rules to protect human subjects today. Because there was a time, and not long ago, where trials were run using unwilling or imprisoned or otherwise disadvantaged subjects.

Well, those are incredibly unethical examples.
Clearly. But sometimes in a randomized trial, there are things you can't get people to give up. There are things scientists

shouldn't ask them to give up. There weren't randomized trials that made people smoke for a decade and not smoke for a decade. It was never done, because it would be unethical.

If you offered to pay people, you probably could get them to eat swordfish three times a week.
Yeah, potentially. But then you also need to check that they are actually eating the way you told them to. Monitoring like that can be expensive. We are talking about studies that would cost billions. And we don't know how long they'd need to last— but we're talking about decades, if you want accurate results.

Not only are the costs of the research already high, but you run into ethical problem of how much to pay the participants. Human investigation committees define a certain level of reimbursement or compensation as coercion. So if you cross the coercion line, they won't let you run the study.

I want to know the small differences of sipping on black coffee versus green tea, snacking on walnuts versus almonds.
The differences are so very small. If you are comparing a good diet to a bad one—oatmeal versus Froot Loops—there's a big difference. If you are comparing the best of the best to the second best of the best — walnuts versus almonds — it's a tiny difference. And one not worth thinking about!

How do we see those tiny differences?
In order to see tiny differences in research, you need huge sample sizes and a lot of time. You have to identify how big a difference you could reasonably expect to see. You now have to look at how many variants—background noise, essentially —you see in real-world populations. One is tempted to say, "Who cares?" with the walnuts-versus-almonds question. But

you do, and you can. But the answer is, "Choose the one you like." Ha!

Do you see these kinds of extensive studies being done in the future? I mean, there is a demand; people are asking these very particular questions about diet.
We don't expect to ever see them done. People want magic, and silver bullets, but neither exists. Walnuts and almonds, both good. Froot Loops and donuts, both bad. We will never nail down every question about diet, but — as we've said all along — we know enough to give and receive really sound advice. And there *is* good news if you want more precise answers.

What's the good news?
There are lesser versions of these expensive, long studies. They are short-term trials — and in particular, short-term randomized controlled trials. Say we want to find out if an optimal vegan diet is better or worse for human health than an optimal vegan diet with fish added — an optimal pescatarian diet. We will randomly assign the subjects to the two groups. Rather than following them for ten years, we will just ask them to play for six months. We will look at a wide array of biomarkers we can capture in that time.

Where can you see a difference in just six months?
Inflammatory markers, shifts in the microbiome, endothelial function. Lipid levels, blood pressure, blood glucose levels. There's a lot of stuff you could get. When reliable biomarkers are used — measures that vary robustly with important outcomes like heart attacks — then measuring these short-term changes becomes a valuable way of projecting what the longer-term, more definitive outcome effects would be. If lipids and inflammatory markers are lower, you can project a lower

rate of heart attack. With lower blood glucose and insulin levels, you can project out less diabetes. And so on. There is a
vast bounty of such studies populating the peer-reviewed literature, and more every day.

**Are there ethics of introducing a new food, like, for
example, trans fat, into a population?**
It's called the precautionary principle. When you are introducing new stuff into a population, the proper thing to do is
assume it's dangerous until you know it's safe. But that is neither a law nor the tradition in the United States: Here, it's introduced and then the population is the guinea pig.

**The precautionary principle presumes guilt until innocence
is proven?**
Exactly. When you are introducing new food to a population,
you should presume you are capable of causing harm until you
prove that you're not. That step was missed with trans fat, and
that step has been missed many times in the history of science
and especially in nutrition. The only guarantees about most
"new" foods is that they won't kill you on the spot, and that
you don't need them.

That's not comforting.
Well, at least science is self-correcting! So over time, food science gave us partially hydrogenated oil. Then, epidemiology
(the study of large populations), another branch of science,
pointed out the harms. And, yes, it was unfortunate. But on
the other hand, in the grand scheme of things, the exposure
was brief. We do think we are getting better at the self-correcting component of science.

**But if science is constantly self-correcting, it seems pretty
easy to shrug science off and say we don't know anything.**

When science does find mistakes it's made, people shouldn't jump to the conclusion that therefore, we shouldn't trust science. We should know how to use science and apply it to diet. And then we should welcome those times that science is self-correcting. It's supposed to do that.

Are there methods other than randomized controlled trials to study diet?

Observation is a powerful source of information. Studying populations as they are is what a focus on randomized controlled trials overlooks. Whether we are talking about putting out campfires or throwing an apple in the air — or, for that matter, smoking and the health effects of smoking. What we know about smoking is from the experience and pattern consistency of large populations, when there is a dramatic change when they smoke. It's not from randomized trials.

So should we just ditch randomized controlled trials altogether? They are expensive, potentially unethical, and subject to human fallibility.

Not at all. What we know most reliably about diet and health, we know from multiple kinds of scientific evidence, including randomized controlled trials. Different kinds of studies fill different gaps in our understanding of what's true. Nothing is better than a randomized controlled trial to help us know attribution with confidence: that a change in Y is attributable to a change in X. Randomized controlled trials are very helpful in establishing what researchers call causality.

So when something is obvious, agreed-upon knowledge, you can skip the scientific research and rely on your own sense.

Yes, PLEASE. Common sense is what is lacking in modern nutrition science. To use our earlier example, our confidence that water quenches thirst and is better than gasoline at putting out fires is so supremely high that we could not possibly expect a formal study of it to clarify our understanding. Science is not the ends, a randomized controlled trial is not the ends, it's the means.

If science is the means, what is "the ends"?

Understanding. The whole point of all of these efforts is to clarify our understanding of what's true. Some things we already know to be true from the monumental consistency of the pattern and from the pattern being so consistent across all the background variations.

But surely it can't hurt to confirm the truth of nutrition with a randomized controlled trial?

The problem is when people believe that everything we know we have to know from a randomized controlled trial — that's fundamentally wrong. It has resulted in the public perception that some experts are undermining the credibility of other experts, and you're forced to choose your champion, choose your hero. Like, "This expert is the *real* expert, those other guys don't know what they are talking about." That's not science. Then people look at the "experts" throwing things at one another's heads and conclude that *none* of them really know what they are talking about, and they tune out.

> You could have a diet that's so good or so bad that the net effect of broccoli would be zilch or nearly zilch.

Knowing how diet is misrepresented in the media will do one of two things: The best-case scenario is that it propagates distrust; the worst-case scenario is that it propagates disgust. Don't renounce nutritional expertise: People believe we know far less about nutrition than we really do; the public is being goaded to toss out the baby with the bathwater. But we actually know a lot!

So even though we don't know everything about the subtleties of diet, we do know quite a lot just from common sense?
And science! But we can say things with much more confidence about overall dietary patterns than about isolated focus. Because frankly, whatever it is you want to say about broccoli, if your overall diet is really terrible, including broccoli wouldn't save you from the effects of that terrible diet. And you could have a really ideal diet without broccoli, too. You could have a diet that's so good or so bad that the net effect of broccoli would be zilch or nearly zilch. We know this! So we can reach far more reliable conclusions about large influences than small dietary patterns. For large influences, the consistency of information on those is really quite incontrovertible.

What do we want to know about nutrition that remains unknown?
There are a lot of important things that we don't know: the effects of adding fish to an optimized vegan diet, what the deal is with dairy, the ketogenic diet . . .

Can we ever know the answers to the questions that are not easily quantified through the scientific method?
Theoretically, we can find the answers to these questions. For example, when it comes to the best diet, as we mentioned before, we don't think anyone would be willing to be randomized for thirty years to be, say, Paleo versus vegan. But we think there are many people who would be willing to participate in a trial that's looking at health effects and weight effects of, let's

say, an optimal vegan diet versus an optimal pescatarian diet versus even a good Paleo diet for a shorter while.

People might sign up for this trial because they are guaranteed to have a really good diet either way.
It's a win-win for the participants and for nutrition science. In this scenario, all options are likely to be better than the participants' baseline diet. If they are overweight, they are gonna lose weight. If they have risk factors, they are likely to improve.

Have we done any scientific experiments on these lifestyle diets?
In the case of both the Mediterranean diet and the vegan diet, there are a number of intervention trials (a trial in which researchers impose some kind of treatment, or intervene) and randomized controlled trials, which tend to answer questions about short-term effects.

If we really want to know what diet is the true winner (which I do), how long would a study have to be?
One hundred years. If we really want to answer what diet is best for health, we would create a randomized controlled trial that assigns people to a dietary pattern at birth and follows them throughout their lives. It hasn't been done, and it will almost certainly never be done.

But think about it like this: Why create a randomized controlled trial when we have all the information we could possibly dream of within these populations of people in the Mediterranean, in Okinawa, in the United States? Drawing information from generations of people is so much more robust than a relatively short-term randomized controlled trial. When we combine what we do know from randomized trials

with what we can only know from observing whole populations over a span of generations, the combination is incredibly
powerful. We know, for example, that the standard American
diet has increased and will increase chronic disease; we don't
need a randomized controlled trial for that.

Conclusion

How to eat well for all the things we care about — health and vitality, weight and appearance, longevity, the well-being of our loved ones, the fate of the planet and, yes, of course, pleasure — is simple. It hides in plain sight. It's the product of copious science, rudimentary sense, and the global consensus of genuine experts. It's the product of heritage and culinary traditions spanning continents, cultures, and generations.

It really is simple; it simply isn't *easy*. Not in our culture.

In our culture today, ultraprocessed food is willfully engineered to be addictive — and everyone is okay with it. Multicolored marshmallows are peddled aggressively to our children as part of a "complete breakfast." Every function of our culture is fueled, fooled, and manipulated by fast food, fizzy drinks, and yes — donuts.

The dietary status quo of our culture is, in a word, revolting. So you should revolt.

You should revolt at every cash register and checkout counter. You should revolt at restaurants. You should revolt at schools, and in conversation with family, friends, and coworkers. You should revolt when you vote.

Our culture is long overdue for a dietary revolution that makes eating well easy rather than elusive. Sure, you can eat well anyway. We hope you do; we hope, by sharing the information in this book, we've helped!

But should you really need to work so hard? It's simple. Our culture should make it easier. So we say: Revolt. We are ready when you are.

Mark will cook.

— David

It seems to me that there are many things to learn from this book, and not all of them fit neatly under the title "How to Eat." That's thanks largely to David, whose mind and knowledge are both broad.

Common sense — what we know to be true (David's usual examples are "apples fall back to the earth" and "don't throw gasoline on fires") — is really important. Groundbreaking, incredible results in scientific studies are rare, and increasingly so. This is worth remembering whether someone claims to have solved for nuclear fusion or determined that — contrary to ten thousand years of experience — wheat is poison.

There's more to take away from this book, of course, because we all have questions about our eating habits. But to me the most fundamental notion, and the most valuable one, is this: We know what good diets are. We absolutely do. Tinkering at the margins — eating more vegetables and fewer whole grains, or vice versa; not eating wheat, or soy, or dairy, or whatever — isn't that important. What's important is choosing real over fake and making sure plants are dominant in your diet. If you've even skimmed this book, you get that.

Common sense would also tell us that a food system that not only allows but helps everyone on earth eat that way is the ideal one. A food system driven by common good would respect the earth itself, us people, and our fellow living creatures. What a concept.

Onward!

— Mark

Select Source Material

PRINCIPAL SOURCE MATERIAL

The True Health Initiative. "Consensus Position." https://www.true-healthinitiative.org/wp-content/uploads/2015/10/OHS_THI_Pledge_160108_FINAL2.pdf.

Oldways Common Ground. "Consensus Statement." https://oldwayspt.org/programs/oldways-common-ground/oldways-common-ground-consensus.

"2015 Dietary Guidelines Advisory Committee Report." https://health.gov/dietaryguidelines/2015-scientific-report/.

Katz, David L. *The Truth About Food*. Amazon White Glove, 2018.

Katz, David L., with Rachel S. C. Friedman and Sean Lucan. *Nutrition in Clinical Practice*. 3rd ed. Philadelphia: Lippincott Williams & Wilkins/Wolters Kluwer, 2014.

Katz, David L., and S. Meller. "Can We Say What Diet Is Best for Health?" *Annu Rev Public Health* 35 (March 18, 2014): 83–103.

US Burden of Disease Collaborators. "The State of US Health, 1990–2016: Burden of Diseases, Injuries, and Risk Factors Among US States." *JAMA* 319, no. 14 (2018): 1444–72. DOI: 10.1001/jama.2018.0158.

GBD 2017 Diet Collaborators. "Health Effects of Dietary Risks in 195 Countries, 1990–2017: A Systematic Analysis for the Global Burden of Disease Study 2017." *Lancet* 393, no. 10184 (May 11, 2019): 1958–72.

Willett, W., J. Rockström, B. Loken, M. Springmann, T. Lang, S. Vermeulen, et al. "Food in the Anthropocene: The EAT-Lancet Commission on Healthy Diets from Sustainable Food Systems." *Lancet* 393, no. 10170 (February 2, 2019): 447–92.

McGinnis, J. M., and W. H. Foege. "Actual Causes of Death in the United States." *JAMA* 270, no. 18 (November 10, 1993): 2207–2212.

Gardner, C. D., J. C. Hartle, R. D. Garrett, L. C. Offringa, and A. S. Wasserman. "Maximizing the Intersection of Human Health and the Health of the Environment with Regard to the Amount and Type of Protein Produced and Consumed in the United States." *Nutr Rev* 77, no. 4 (April 1, 2019): 197–215.

Katz, D. L., K. N. Doughty, K. Geagan, D. A. Jenkins, and C. D. Gardner. "Perspective: The Public Health Case for Modernizing the Definition of Protein Quality." *Adv Nutr* 10, no. 5 (September 1, 2019): 755–64.

Nestle, Marion. *Food Politics*. Oakland: University of California Press, 2007.

Pollan, Michael. *In Defense of Food*. New York: Penguin Books, 2009.

Moss, Michael. *Salt Sugar Fat*. New York: Random House, 2014.

Buettner, Dan. *The Blue Zones*. Washington, DC: National Geographic, 2010.

Foodtank, https://foodtank.com/.

EAT Forum, https://eatforum.org/.

Planetary Health Alliance, https://planetaryhealthalliance.org/.

Civil Eats, https://civileats.com/.

Center for Science in the Public Interest, https://cspinet.org/.

ADDITIONAL SOURCE MATERIAL

Akesson, A., S. C. Larsson, A. Discacciati, and A. Wolk. "Low-Risk Diet and Lifestyle Habits in the Primary Prevention of Myocardial Infarction in Men: A Population-Based Prospective Cohort Study." *J Am Coll Cardiol* 64, no. 13 (September 30, 2014): 1299–306.

Akesson, A., C. Weismayer, P. K. Newby, and A. Wolk. "Combined Effect of Low-Risk Dietary and Lifestyle Behaviors in Primary Prevention of Myocardial Infarction in Women." *Arch Intern Med* 167, no. 19 (October 22, 2007): 2122–7.

Aldana, S. G., R. L. Greenlaw, H. A. Diehl, A. Salberg, R. M. Merrill, S. Ohmine, and C. Thomas. "The Behavioral and Clinical Effects of Therapeutic Lifestyle Change on Middle-Aged Adults." *Prev Chronic Dis* 3, no. 1 (January 2006): A05.

Aleksandrova, K., et al. "Combined Impact of Healthy Lifestyle Factors on Colorectal Cancer: A Large European Cohort Study." *BMC Med* 12, no. 1 (October 10, 2014): 168. [Epub ahead of print]

Allen, N. B., L. Zhao, L. Liu, M. Daviglus, K. Liu, J. Fries, Y. T. Shih, et al. "Favorable Cardiovascular Health, Compression of Morbidity, and Healthcare Costs: Forty-Year Follow-Up of the CHA Study (Chicago Heart Association Detection Project in Industry)." *Circulation* 135, no. 18 (May 2, 2017): 1693–701.

Chiuve, S. E., K. M. Rexrode, D. Spiegelman, G. Logroscino, J. E. Manson, and E. B. Rimm. "Primary Prevention of Stroke by Healthy Lifestyle." *Circulation* 118, no. 9 (August 26, 2008): 947–54.

Chomistek, A. K., S. E. Chiuve, A. H. Eliassen, K. J. Mukamal, W. C. Willett, and E. B. Rimm. "Healthy Lifestyle in the Primordial Prevention of Cardiovascular Disease Among Young Women." *J Am Coll Cardiol* 65, no. 1 (January 6, 2015): 43–51.

Daar, A. S., P. A. Singer, D. L. Persad, S. K. Pramming, D. R. Matthews, R. Beaglehole, A. Bernstein, et al. "Grand Challenges in Chronic Non-Communicable Diseases." *Nature* 450, no. 7169 (November 22, 2007): 494–6.

Dansinger, M. L., J. A. Gleason, J. L. Griffith, H. P. Selker, and E. J. Schaefer. "Comparison of the Atkins, Ornish, Weight Watchers, and Zone Diets for Weight Loss and Heart Disease Risk Reduction: A Randomized Trial." *JAMA* 293, no. 1 (January 5, 2005): 43–53.

De Lorgeril, M., P. Salen, J. L. Martin, N. Mamelle, I. Monjaud, P. Touboul, and J. Delaye. "Effect of a Mediterranean Type of Diet on the Rate of Cardiovascular Complications in Patients with Coronary Artery Disease. Insights into the Cardioprotective Effect of Certain Nutriments." *J Am Coll Cardiol* 28, no. 5 (November 1, 1996): 1103–8.

De Waure, C., G. J. Lauret, W. Ricciardi, B. Ferket, J. Teijink, S. Spronk, and M. G. Myriam Hunink. "Lifestyle Interventions in Patients with Coronary Heart Disease: A Systematic Review." *Am J Prev Med* 45, no. 2 (August 2013): 207–16.

Estruch, R., E. Ros, J. Salas-Salvadó, M. I. Covas, D. Corella, F. Arós, E. Gómez-Gracia, et al.; PREDIMED Study Investigators. "Primary Prevention of Cardiovascular Disease with a Mediterranean Diet." *N Engl J Med* 368, no. 14 (April 4, 2013): 1279–90.

Ford, E. S., M. M. Bergmann, J. Kröger, A. Schienkiewitz, C. Weikert, and H. Boeing. "Healthy Living Is the Best Revenge: Findings from the European Prospective Investigation into Cancer and Nutrition-Potsdam Study." *Arch Intern Med* 169, no. 15 (August 10, 2009): 1355–62.

Freeman, A. M., P. B. Morris, N. Barnard, C. B. Esselstyn, E. Ros, A. Agatston, S. Devries, et al. "Trending Cardiovascular Nutrition Controversies." *J Am Coll Cardiol* 69, no. 9 (March 7, 2017): 1172–87.

Galimanis, A., M. L. Mono, M. Arnold, K. Nedeltchev, and H. P. Mattle. "Lifestyle and Stroke Risk: A Review." *Curr Opin Neurol* 22, no. 1 (February 2009): 60–68.

Gardner, C. D., A. Kiazand, S. Alhassan, S. Kim, R. S. Stafford, R. R. Balise, H. C. Kraemer, and A. C. King. "Comparison of the Atkins, Zone, Ornish, and LEARN Diets for Change in Weight and Related Risk Factors Among Overweight Premenopausal Women: The A TO Z Weight Loss Study: A Randomized Trial." *JAMA* 297, no. 9 (March 7, 2007): 969–77.

Gardner, C. D. "Tailoring Dietary Approaches for Weight Loss." *Int J Obes* 2, suppl. 1 (July 2012): S11–S15.

Gopinath, B., E. Rochtchina, V. M. Flood, and P. Mitchell. "Healthy Living and Risk of Major Chronic Diseases in an Older Population." *Arch Intern Med* 170, no. 2 (January 25, 2010): 208–9.

Gregg, E. W., H. Chen, L. E. Wagenknecht, J. M. Clark, L. M. Delahanty,

J. Bantle, H. J. Pownall, et al.; Look AHEAD Research Group. "Association of an Intensive Lifestyle Intervention with Remission of Type 2 Diabetes." *JAMA* 308, no. 23 (December 19, 2012): 2489–96.

Gupta, B. P., M. H. Murad, M. M. Clifton, L. Prokop, A. Nehra, and S. L. Kopecky. "The Effect of Lifestyle Modification and Cardiovascular Risk Factor Reduction on Erectile Dysfunction: A Systematic Review and Meta-Analysis." *Arch Intern Med* 171, no. 20 (November 14, 2011): 1797–803.

Holme, I., K. Retterstøl, K. R. Norum, and I. Hjermann. "Lifelong Benefits on Myocardial Infarction Mortality: 40-Year Follow-Up of the Randomized Oslo Diet and Antismoking Study." *J Intern Med* 280, no. 2 (August 2016): 221–7.

Jenkins, D. J., B. A. Boucher, F. D. Ashbury, M. Sloan, P. Brown, A. El-Sohemy, A. J. Hanley, et al. "Effect of Current Dietary Recommendations on Weight Loss and Cardiovascular Risk Factors." *J Am Coll Cardiol* 69, no. 9 (March 7, 2017): 1103–12.

Jenkins, D. J., P. J. Jones, B. Lamarche, C. W. Kendall, D. Faulkner, L. Cermakova, I. Gigleux, et al. "Effect of a Dietary Portfolio of Cholesterol-Lowering Foods Given at 2 Levels of Intensity of Dietary Advice on Serum Lipids in Hyperlipidemia: A Randomized Controlled Trial." *JAMA* 306, no. 8 (August 24, 2011): 831–9.

Jenkins, D. J., A. R. Josse, J. M. Wong, T. H. Nguyen, and C. W. Kendall. "The Portfolio Diet for Cardiovascular Risk Reduction." *Curr Atheroscler Rep* 9, no. 6 (December 2007): 501–7.

Jousilahti, P., T. Laatikainen, M. Peltonen, K. Borodulin, S. Männistö, A. Jula, V. Salomaa, K. Harald, P. Puska, and E. Vartiainen. "Primary Prevention and Risk Factor Reduction in Coronary Heart Disease Mortality among Working Aged Men and Women in Eastern Finland Over 40 Years: Population Based Observational Study." *BMJ* 352 (March 1, 2016): i721.

Katz, D. L., E. P. Frates, J. P. Bonnet, S. K. Gupta, E. Vartiainen, and R. H. Carmona. "Lifestyle as Medicine: The Case for a True Health Initiative." *Am J Health Promot* (January 1, 2017). DOI: 10.1177/0890117117705949. [Epub ahead of print]

Katz, D. L., and F. B. Hu. "Knowing What to Eat, Refusing to Swallow It." *Huffington Post,* July 2, 2014.

Katz, D. L. "Lifestyle Is the Medicine, Culture Is the Spoon: The Covariance of Proposition and Preposition." *Am J Lifestyle* Med 8 (2014): 301–5.

Katz, D. L. "Life and Death, Knowledge and Power: Why Knowing What Matters Is Not What's the Matter." *Arch Intern Med* 169, no. 15 (August 10, 2009): 1362–3.

Khera, A. V., C. A. Emdin, I. Drake, P. Natarajan, A. G. Bick, N. R. Cook, D. I. Chasman, et al. "Genetic Risk, Adherence to a Healthy Lifestyle, and Coronary Disease." *N Engl J Med* 375, no. 24 (December 15, 2016): 2349–58.

King, D. E., A. G. Mainous 3rd, M. Carnemolla, and C. J. Everett. "Adherence to Healthy Lifestyle Habits in US Adults, 1988–2006." *Am J Med* 122, no. 6 (June 2009): 528–34.

Knoops, K. T., L. C. de Groot, D. Kromhout, A. E. Perrin, O. Moreiras-Varela, A. Menotti, and W. A. van Staveren. "Mediterranean Diet, Lifestyle Factors, and 10-Year Mortality in Elderly European Men and Women: The HALE Project." *JAMA* 292, no. 12 (September 22, 2004): 1433–9.

Knowler, W. C., E. Barrett-Connor, S. E. Fowler, R. F. Hamman, J. M. Lachin, E. A. Walker, and D. M. Nathan; Diabetes Prevention Program Research Group. "Reduction in the Incidence of Type 2 Diabetes with Lifestyle Intervention or Metformin." *N Engl J Med* 346, no. 6 (February 7, 2002): 393–403.

Kono, Y., S. Yamada, J. Yamaguchi, Y. Hagiwara, N. Iritani, S. Ishida, A. Araki, Y. Hasegawa, H. Sakakibara, and Y. Koike. "Secondary Prevention of New Vascular Events with Lifestyle Intervention in Patients with Noncardioembolic Mild Ischemic Stroke: A Single-Center Randomized Controlled Trial." *Cerebrovasc Dis* 36, no. 2 (2013): 88–97.

Kurth, T., S. C. Moore, J. M. Gaziano, C. S. Kase, M. J. Stampfer, K. Berger, and J. E. Buring. "Healthy Lifestyle and the Risk of Stroke in Women." *Arch Intern Med* 166, no. 13 (July 10, 2006): 1403–9.

Kvaavik, E., G. D. Batty, G. Ursin, R. Huxley, and C. R. Gale. "Influence of Individual and Combined Health Behaviors on Total and Cause-Specific Mortality in Men and Women: The United Kingdom Health and Lifestyle Survey." *Arch Intern Med* 170, no. 8 (April 26, 2010): 711–8.

Ley, S. H., O. Hamdy, V. Mohan, and F. B. Hu. "Prevention and Management of Type 2 Diabetes: Dietary Components and Nutritional Strategies." *Lancet* 383 (June 7, 2014): 1999–2007.

Li, Y., A. Hruby, A. M. Bernstein, S. H. Ley, D. D. Wang, S. E. Chiuve, L. Sampson, K. M. Rexrode, E. B. Rimm, W. C. Willett, and F. B. Hu. "Saturated Fats Compared with Unsaturated Fats and Sources of Carbohydrates in Relation to Risk of Coronary Heart Disease: A Prospective Cohort Study." *J Am Coll Cardiol* 66, no. 14 (October 6, 2015): 1538–48.

Loef, M., and H. Walach. "The Combined Effects of Healthy Lifestyle Behaviors on All Cause Mortality: A Systematic Review and Meta-Analysis." *Prev Med* 55, no. 3 (September 2012): 163–70.

Machovina, B., K. J. Feeley, and W. J. Ripple. "Biodiversity Conservation: The Key Is Reducing Meat Consumption." *Sci Total Environ* 536 (December 1, 2015): 419–31.

Mann, J., et al. "Low Carbohydrate Diets: Going Against the Grain." *Lancet* 384 (October 25, 2014): 1479–80.

McCullough, M. L., A. V. Patel, L. H. Kushi, R. Patel, W. C. Willett, C. Doyle, M. J. Thun, and S. M. Gapstur. "Following Cancer Prevention Guidelines Reduces Risk of Cancer, Cardiovascular Disease, and All-Cause Mortality." *Cancer Epidemiol Biomarkers Prev* 20, no. 6 (June 2011): 1089–97.

Meng, L., G. Maskarinec, J. Lee, and L. N. Kolonel. "Lifestyle Factors and Chronic Diseases: Application of a Composite Risk Index." *Prev Med* 29, no. 4 (October 1999): 296–304.

Menotti, A., D. Kromhout, P. E. Puddu, A. Alberti-Fidanza, P. Hollman, A. Kafatos, H. Tolonen, H. Adachi, and D. R. Jacobs Jr. "Baseline Fatty Acids, Food Groups, a Diet Score and 50-Year All-Cause Mortality Rates. An Ecological Analysis of the Seven Countries Study." *Ann Med* (September 6, 2017): 1–10. DOI: 10.1080/07853890.2017.1372622.

Micha, R., J. L. Peñalvo, F. Cudhea, F. Imamura, C. D. Rehm, and D. Mozaffarian. "Association Between Dietary Factors and Mortality from Heart Disease, Stroke, and Type 2 Diabetes in the United States." *JAMA* 317, no. 9 (March 7, 2017): 912–24.

Mokdad, A. H., J. S. Marks, D. F. Stroup, and J. L. Gerberding. "Actual Causes of Death in the United States, 2000." *JAMA* 291, no. 10 (March 10, 2004): 1238–45.

Mozaffarian, D. "Dietary and Policy Priorities for Cardiovascular Disease, Diabetes, and Obesity: A Comprehensive Review." *Circulation* 133, no. 2 (January 12, 2016): 187–225.

Muchiteni, T, and W. B. Borden. "Improving Risk Factor Modification: A Global Approach." *Curr Cardiol Rep* 11, no. 6 (November 2009): 476–83.

Ngandu, T., J. Lehtisalo, A. Solomon, E. Levälahti, S. Ahtiluoto, R. Antikainen, L. Bäckman, et al. "A 2-Year Multidomain Intervention of Diet, Exercise, Cognitive Training, and Vascular Risk Monitoring versus Control to Prevent Cognitive Decline in At-Risk Elderly People (FINGER): A Randomised Controlled Trial." *Lancet* 385, no. 9984 (June 6, 2015): 2255–63.

Nicklett, E. J., R. D. Semba, Q. L. Xue, J. Tian, K. Sun, A. R. Cappola, E. M. Simonsick, L. Ferrucci, and L. P. Fried. "Fruit and Vegetable Intake, Physical Activity, and Mortality in Older Community-Dwelling Women." *J Am Geriatr Soc* 60, no. 5 (May 2012): 862–8.

Ornish, D., J. Lin, J. Daubenmier, G. Weidner, E. Epel, C. Kemp, M. J. Magbanua, R. Marlin, L. Yglecias, P. R. Carroll, and E. H. Blackburn. "Increased Telomerase Activity and Comprehensive Lifestyle Changes: A Pilot Study." *Lancet Oncol* 9, no. 11 (November 2008): 1048–57.

Ornish, D., M. J. Magbanua, G. Weidner, V. Weinberg, C. Kemp, C. Green, M. D. Mattie, et al. "Changes in Prostate Gene Expression in Men Undergoing an Intensive Nutrition and Lifestyle Intervention." *Proc Natl Acad Sci U S A* 105, no. 24 (June 17, 2008): 8369–74.

Ornish, D., L. W. Scherwitz, J. H. Billings, S. E. Brown, K. L. Gould, T. A. Merritt, S. Sparler, et al. "Intensive Lifestyle Changes for Reversal of Coronary Heart Disease." *JAMA* 280, no. 23 (December 16, 1998): 2001–7.

Pett, K. D., J. Kahn, W. C. Willett, and D. L. Katz. "Ancel Keys and the Seven Countries Study: An Evidence-Based Response to Revisionist Histories." *True Health Initiative.* http://www.truehealthinitiative.org/wordpress/wp-content/uploads/2017/07/SCS-White-Paper.THI_.8-1-17.pdf.

Ramsey, F., A. Ussery-Hall, D. Garcia, G. McDonald, A. Easton, M. Kambon, L. Balluz, W. Garvin, and J. Vigeant; Centers for Disease Control and Prevention (CDC). "Prevalence of Selected Risk Behaviors and Chronic Diseases — Behavioral Risk Factor Surveillance System (BRFSS), 39 Steps Communities, United States, 2005." *MMWR Surveill Summ* 57, no. 11 (October 31, 2008): 1–20.

Schellenberg, E. S., D. M. Dryden, B. Vandermeer, C. Ha, and C. Korownyk. "Lifestyle Interventions for Patients with and at Risk for Type 2 Diabetes: A Systematic Review and Meta-Analysis." *Ann Intern Med* 159, no. 8 (October 15, 2013): 543–51.

Small, B. J., R. A. Dixon, J. J. McArdle, and K. J. Grimm. "Do Changes in Lifestyle Engagement Moderate Cognitive Decline in Normal Aging? Evidence from the Victoria Longitudinal Study." *Neuropsychology* 26, no. 2 (March 2012): 144–55.

Sofi, F., M. Dinu, G. Pagliai, F. Cesari, R. Marcucci, and A. Casini. "Mediterranean versus Vegetarian Diet for Cardiovascular Disease Prevention (the CARDIVEG Study): Study Protocol for a Randomized Controlled Trial." *Trials* 17, no. 1 (May 4, 2016): 233.

Song, M., T. T. Fung, F. B. Hu, W. C. Willett, V. D. Longo, A. T. Chan, and E. L. Giovannucci. "Association of Animal and Plant Protein Intake with All-Cause and Cause-Specific Mortality." *JAMA Intern Med* (August 1, 2016). DOI: 10.1001/jamainternmed.2016.4182.

Song, M., and E. Giovannucci. "Preventable Incidence and Mortality of Carcinoma Associated with Lifestyle Factors among White Adults in the United States." *JAMA Oncol* 2, no. 9 (September 1, 2016): 1154–61.

Sotos-Prieto, M., S. N. Bhupathiraju, J. Mattei, T. T. Fung, Y. Li, A. Pan, W. C. Willett, E. B. Rimm, and F. B. Hu. "Association of Changes in Diet Quality with Total and Cause-Specific Mortality." *N Engl J Med* 377, no. 2 (July 13, 2017): 143–53.

Spencer, E. A., K. L. Pirie, R. J. Stevens, V. Beral, A. Brown, B. Liu, J. Green, and G. K. Reeves; Million Women Study Collaborators. "Diabetes and

Modifiable Risk Factors for Cardiovascular Disease: The Prospective Million Women Study." *Eur J Epidemiol* 23, no. 12 (2008): 793–9.

Springmann, M., H. C. Godfray, M. Rayner, and P. Scarborough. "Analysis and Valuation of the Health and Climate Change Cobenefits of Dietary Change." *Proc Natl Acad Sci U S A* 113, no. 15 (April 12, 2016): 4146–51.

Steptoe, A., and J. Wardle. "What the Experts Think: A European Survey of Expert Opinion About the Influence of Lifestyle on Health." *Eur J Epidemiol* 10, no. 2 (April 1994): 195–203.

Stewart, B. W. "Priorities for Cancer Prevention: Lifestyle Choices versus Unavoidable Exposures." *Lancet Oncol* 13, no. 3 (March 2012): e126–33.

Tanaka, S., S. Yamamoto, M. Inoue, M. Iwasaki, S. Sasazuki, H. Iso, and S. Tsugane; JPHC Study Group. "Projecting the Probability of Survival Free from Cancer and Cardiovascular Incidence Through Lifestyle Modification in Japan." *Prev Med* 48, no. 2 (February 2009): 128–33.

Trichopoulou, A., C. Bamia, and D. Trichopoulos. "Anatomy of Health Effects of Mediterranean Diet: Greek EPIC Prospective Cohort Study." *BMJ* 338 (June 23, 2009): b2337. DOI: 10.1136/bmj.b2337.

Turner-McGrievy, G. M., C. R. Davidson, E. E. Wingard, S. Wilcox, and E. A. Frongillo. "Comparative Effectiveness of Plant-Based Diets for Weight Loss: A Randomized Controlled Trial of Five Different Diets." *Nutrition* 31, no. 2 (February 2015): 350–8.

Wang, D. D., Y. Li, S. E. Chiuve, M. J. Stampfer, J. E. Manson, E. B. Rimm, W. C. Willett, and F. B. Hu. "Association of Specific Dietary Fats with Total and Cause-Specific Mortality." *JAMA Intern Med* 176, no. 8 (August 1, 2016): 1134–45.

Wannamethee, S. G., A. G. Shaper, M. Walker, and S. Ebrahim. "Lifestyle and 15-Year Survival Free of Heart Attack, Stroke, and Diabetes in Middle-Aged British Men." *Arch Intern Med* 158, no. 22 (December 7–21, 1998): 2433–40.

Weisburger, J. H. "Lifestyle, Health and Disease Prevention: The Underlying Mechanisms." *Eur J Cancer Prev* 11, suppl. 2 (August 2002): S1–7.

Woo, J. "Relationships among Diet, Physical Activity and Other Lifestyle Factors and Debilitating Diseases in the Elderly." *Eur J Clin Nutr* 54, suppl. 3 (June 2000): S143–7.

Index